Waste Minimisation: A Chemist's Approach

Waste Minimisation:
A Chemist's Approach

Edited by

K. Martin

Contract Catalysts, a division of
Contract Chemicals (Knowsley) Ltd, Prescot, Merseyside

T. W. Bastock

Contract Chemicals (Knowsley) Ltd, Prescot, Merseyside

ROYAL
SOCIETY OF
CHEMISTRY

The Proceedings of a Joint Symposium entitled 'The Chemist's Contribution to Waste Minimisation' held by the Royal Society of Chemistry and the Society of Chemical Industry at the University of Lancaster, UK, on 30 June–1 July 1993.

Special Publication No. 140

ISBN 0-85186-585-2

A catalogue record for this book is available from the British Library

© The Royal Society of Chemistry 1994

Published by The Royal Society of Chemistry,
Thomas Graham House, The Science Park, Milton Road,
Cambridge CB4 4WF

Printed in Great Britain by Bookcraft (Bath) Ltd

Preface

The Chemical Industry produces a vast array of products which are an integral part of everyday life. The products make an enormous contribution to our comfort and well-being. However, these manufacturing processes lead to millions of tonnes of waste comprising liquid and solid effluent, atmospheric emissions, and packaging waste. The elimination of these wastes or at least their safe, clean disposal has become a major issue in the developed world.

The increasing awareness of the general public on environmental issues has brought pressure on government bodies to introduce legislation that will conserve and protect the environment. In the UK the Environmental Protection Act (EPA), enforced and policed by Her Majesty's Inspectorate of Pollution (HMIP), has far reaching powers and affects all aspects of chemistry in industry and academia. Further, industry self-regulation through Responsible Care initiatives has accelerated the pace of process improvement and reduction of environmental impact.

Waste Minimisation: A Chemist's Approach is the result of a conference that was held at Lancaster University on 30th June - 1st July 1993. The lectures given at this conference are reproduced in this book. The contents cover all aspects of waste minimisation techniques, aiming especially at the elimination at source rather than 'end of pipe' solutions. The current legal requirements of the EPA together with views on future legislation are included. The collaboration of Government, Industry, and Academia in the finding and implementation of novel, clean technology is of particular importance.

The contributions have been arranged into four sections and an attempt has been made to link these sections together with a progressive theme:

Section One - Legal Aspects of Waste Minimisation - looks at government policies in general and then reviews the role of HMIP. It defines and interprets the best Practicable Environmental Option (BPEO) and Best Available Techniques Not Entailing Excessive Cost (BATNEEC). This is followed by the CIA view to waste minimisation and clean technology.

Section Two - Corporate Strategy to Waste Minimisation - is a description of how four companies (BASF, Shell, Monsanto, and Contract Chemicals) are responding to the environmental legislation.

Section Three - Practical Aspects of Waste Minimisation - describes how effluent can be reduced. In particular this includes 'End of Pipe' techniques, which can give dramatic reductions in waste, and the use of recycling at every opportunity, which not only reduces the effluent produced but also saves money making a process more cost effective.

Section Four - New Technology for Waste Minimisation - takes the approach of 'prevention is better than cure'. It deals with new catalysts for existing processes which produce little or no effluent and computer monitored hardware that can be used to efficiently optimise a process to give maximum yield with minimum waste.

In conclusion I would like to thank all the contributors to this book for their efforts in converting their presentations into papers for use in this publication.

Keith Martin
September, 1993

Contents

Legal Aspects of Waste Minimisation

Government Policies to Encourage Cleaner Production

Lord Strathclyde

PARLIAMENTARY UNDER SECRETARY OF STATE, DEPARTMENT OF THE
ENVIRONMENT, 2 MARSHAM STREET, LONDON SW1P 3EB, UK

Professor Mansfield, Ladies, and Gentlemen, I am very
pleased and honoured to have been invited to address this
opening session of the Royal Society of Chemistry's
Symposium, "The Chemist's Contribution to Waste
Minimisation". I will be followed by Dr David Slater,
Her Majesty's Chief Inspector of Pollution, who will
speak on "Waste Minimisation in the context of Integrated
Pollution Control". Before I go on, perhaps I should say
what I mean by "cleaner production". I hope it turns out
that it includes something similar to what the RSC - and
David Slater - mean by "waste minimisation". Quite a few
phrases have been coined over the years on this theme.
In their report last year entitled "Cleaner Technology",
ACOST, the Advisory Council on Science and Technology,
included a whole annex several pages long solely devoted
to definitions.

Definitions

When I refer to "cleaner production" I am happy to
take a lead from UNEP, the United Nations Environment
Programme. UNEP define cleaner production as involving
"the continuous application of an integrated preventive
environmental strategy to processes and products to
reduce risks to humans and the environment". They
amplify this by saying that, for production processes,
"cleaner production includes conserving raw materials and
energy, eliminating toxic raw materials, and reducing the
quantity and toxicity of all emissions and wastes before
they leave a process." The goal of cleaner production is
"to avoid generating waste in the first place and to
minimise the use of raw materials and energy".

I think you will agree that cleaner production, so
defined, embraces what is often meant by waste
minimisation. It goes somewhat further in relation to
life cycle analysis of products. The expressions
"cleaner production" and "waste minimisation" have
important connotations of reconciling three highly

important but at first sight conflicting objectives for the UK and the whole world: economic growth, conservation of resources and protection of the environment.

It may sometimes be thought - quite incorrectly - that my Department is concerned with the second two objectives to the exclusion of the first. I believe it is highly significant that my Department last year took on the Energy Efficiency Office and its work from the old Department of Energy. The EEO has been striving for many years to improve the efficiency with which industry uses a key resource - energy. I regard my Department as having an important responsibility to promote, not only energy efficiency but what might be called **environmental** efficiency. By that I mean encouraging industry to adopt strategies for meeting - **and surpassing** - environmental standards in the **most efficient** manner possible. Such strategies will involve the most efficient use of raw materials and therefore the least waste from the production process, so that both the economy and the environment are winners.

Policy

So what are the policies and policy instruments available to the Government to promote cleaner production (or waste minimisation or environmental efficiency)? The key **policy** is represented by the hierarchy of waste management options set out, for example, in a European Community Council Resolution on Waste Policy in 1990. The hierarchy places complete elimination of waste at the top of the list. It then proceeds via reduction at source, through recycling to treatment and disposal. You will probably recognise that this hierarchy is embodied in the legislation on Integrated Pollution Control, on which David Slater will be speaking in a moment.

Instruments

Therefore a major **instrument** at Government's disposal for pursuing cleaner production is the traditional one of legislation and regulation. There are however other means. It is important, if we are to be consistent with the Government's overall policy of minimising the extent and complexity of regulation, that we should look carefully at how far our objectives can be pursued by other means. I would like to touch briefly on non-regulatory instruments such as economic measures; the development of management procedures; information on best practice; grant support and environmental awards.

Regulation

Before I go on to non-regulatory areas, however, I would like to touch on another area of legislation which complements Integrated Pollution Control and allied systems of regulation. This is waste management legislation. The legislation establishing a "duty of care", the reform of waste disposal and regulation authorities and the new licensing system for waste disposal sites will all result in enhanced environmental protection at the level of waste handling and disposal. It will also, inevitably and some would say <u>desirably</u>, increase the costs to industry of waste disposal and therefore most importantly provide an incentive for industry to minimise the amount of waste which arises.

Pressure on industry through the withdrawal of traditional waste disposal avenues can often produce beneficial results. I would quote as an example the new ICI Monomer 8 Plant at Billingham which Michael Howard formally opened in April 1993 while he was still at the DOE. In that case, ICI faced a situation in which a traditional approach to manufacture of methyl methacrylate would have created a severe waste disposal problem following the banning of the established disposal route, which involved discharging many thousands of tonnes of contaminated acid to the North Sea. Our treaty obligations arising from the North Sea Conference meant that sea disposal had to end. However, ICI were not simply content with finding a more acceptable waste disposal method. Instead they completely redesigned their process, installing the world's largest and most advanced sulphuric acid recovery plant. Material which was once regarded as a worthless waste is now being recycled back into the process to the benefit of all concerned.

Economic Instruments

"Traditional" legislation does of course have economic effects but we are committed to investigating what scope there may be for adopting more direct economic instruments in pursuit of environmental objectives. In the UK, the most familiar example of an economic instrument used to secure an environmental benefit is still the case of unleaded petrol. A little closer to the theme of this conference, we have also introduced, from April 1992, a system of <u>recycling</u> <u>credits</u>. This allows organisations which have saved the waste disposal authorities money by recycling household waste (which would otherwise have become their responsibility) to claim cash payments reflecting a proportion of the savings. The main beneficiaries of the system are waste collection authorities but community groups can also benefit. For example, Dorset County Council have over forty community bodies - schools, churches, scout groups, sports clubs and so on - on their recycling credit register. These bodies can, in Dorset's case, claim £4.62 for each tonne of material they collect - which is of course additional to whatever they have received from

sale of the material.   One scout group in Dorset has,
over the last year, collected over 145 tonnes of paper
and earned over £650 in credits.

Work is ongoing on the question of charging for
liquid effluent releases and on tradeable permits for SO_2
emissions.   The Government has recently published two
reports on landfill costs and prices, and on economic
instruments and recovery of resources from waste.
Further work is now in hand on the environmental impacts
of landfill and incineration. Properly applied, economic
instruments could make waste producers face the full
costs of waste disposal and any associated environmental
costs.   This would give a further push in the direction
of waste minimisation and recycling.

Management Procedures

A company that is concerned about its environmental
performance must first ensure that its top management is
prepared to be committed to taking all reasonable steps
to making improvements wherever feasible.   The company
should have an explicit environmental policy.   The next
step should be to conduct a systematic review of the
environmental impacts at its operational sites.   I can
increasingly foresee the production of a convincing
environmental review becoming a vital management tool for
any responsible company concerned about its corporate
image, its credit rating and customer acceptance of its
products.

The Government has supported the development of
British Standard BS 7750 for environmental management
systems.   This is a European first and is creating a
great deal of interest overseas as well as at home.   Any
company wishing to develop an environmental management
system which is certified according to BS 7750 will have
to adopt a range of procedures which are intended to
improve its environmental performance as an ongoing
process.   These will have to include measures which
relate to waste minimisation and other waste management
issues.   The pilot testing of BS 7750 has now been
completed and the final version should be published by
the end of 1993.

The UK has also been very active in the development
of the European Community Eco-Management and Audit
Regulation which has just been adopted by the Council of
Ministers.   We envisage that a company certified to BS
7750 will have fulfilled most of the requirements of the
EC scheme, which, like BS 7750, is voluntary.   The only
significant extra task to gain registration and Europe-
wide recognition under the Regulation will be to publish
an independently verified environmental statement.   Now
that the Regulation has been adopted, we have 21 months
to set up UK arrangements for the scheme.

We intend to put out a consultation paper quite soon to indicate our preferred options for setting up the "competent body" to administer the scheme in the UK. We also need to identify the organisation which will accredit and supervise those qualified to verify the environmental statements.

BS 7750 and the EC Eco-Audit Regulation are very general procedures. Closer to the theme of this particular symposium, I would like to mention another useful initiative which my Department was able to undertake. This was the commissioning through the Institution of Chemical Engineers of a draft Waste Minimisation Guide, drawing on the pioneering work of the US Environmental Protection Agency. The draft Guide is being tried out by, for example, the participants in the Aire & Calder waste minimisation demonstration programme launched last year with support from Government and industry in West Yorkshire. As HMIP is directly involved in that project, I'll leave Dr Slater to say more about it.

Information

I would however like to develop and generalise the reference to the Aire & Calder project. One of the other areas in which I believe the Government can potentially make a very valuable contribution to the pursuit of cleaner production and waste minimisation is the dissemination in various ways of information about existing best practice. There is plenty to suggest that if every firm in a given sector did as well in pollution prevention and control as the best in that sector, then very substantial improvements could be achieved. This may often be largely a matter of good housekeeping with little or no need for capital investment. But there may also be ample scope in many cases for companies to undertake investment in process modifications which will reduce pollution and yield efficiency improvements sufficient to pay back the investment in what may be a surprisingly short time. Both my Department and the Department of Trade and Industry have produced publications containing case histories to illustrate this crucial point. Most recently, I was very pleased that my Department was able to produce for UNEP a booklet called "Cleaner Production Worldwide". This is intended to encourage decision makers in industry around the world to consider the scope for reduction of pollution at source in their own operations.

I firmly believe that a few well selected, concisely presented demonstration projects are more likely to engage the attention of busy industrial managers than more generalised or technical material - necessary though that also is. The DTI have run for the last three years a demonstration project scheme - DEMOS. This is no longer open to new cases but it has produced a number of very interesting projects, some of which are still at an early stage. One such project is "Catalyst".

Like the Aire & Calder project, the Catalyst project brings together a number of firms, in this case fifteen based on Merseyside. The object is to see what reductions could be achieved in emissions and waste of all kinds by carrying out comprehensive audits and drawing up waste reduction plans as a result. The Catalyst project was initiated by my own Department's Environmental Technology Adviser, although I am happy to acknowledge that the Government funding is coming wholly, in this case, from the DTI's DEMOS scheme. There is equal very generous financial support from the BOC Foundation for the Environment.

Catalyst was officially launched by my colleague Neil Hamilton, Corporate Affairs Minister at DTI in June 1993. Several of the participating companies are in the chemical sector. Information about Catalyst events and eventually on results can be obtained from the project consultants, W S Atkins at Warrington.

Grants

My Department, and the DTI, have also tried to help stimulate progress in new environmental technology by means of grants for pre-competitive R & D. From my own Department's point of view, the rationale has been the belief that in some areas progress in tightening environmental standards relies on improved technology. The Environmental Technology Innovation Scheme, ETIS, which DOE launched jointly with DTI in 1990, was originally planned to remain open to new applications until autumn 1993. Resource constraints have meant that the Departments have already had to close the scheme to new applications. Nevertheless we have been able to assist more than forty interesting projects, and we may be able to make a small number of further grant offers to applications currently in our system. The Departments are now considering what form a new environmental technology support scheme might best take (if the necessary resources can be secured), and what its priorities should be. Our first impression is that the overall priority should be encouraging fuller and wider use of existing techniques and technology. This will build on our previous work on case history publications and on projects such as Aire & Calder and Catalyst. We are looking at possible analogies with the Best Practice programme of the Energy Efficiency Office and other models. We would welcome views from anyone here as to how they think a new "Environmental Best Practice Programme" might best be constructed and prioritised.

Awards

An effective and appropriate way of drawing attention to success stories in environmental technology and environmental management is the presentation of prestigious Awards. For several years, DOE supported the Royal Society of Arts' Better Environment Awards for Industry scheme. The last of these awards were presented in 1992. One of the awards went to the University of York and Contract Chemicals for their work on environmentally friendly catalysts. I know that Dr Bastock of Contract Chemicals has been closely involved in the organisation of this event. I also see from your programme that you will be having a presentation tomorrow from Dr Clark of the University of York. The RSA discontinued the BEAFI scheme for the excellent reason that their work has been taken up by the Queen's Awards system. In April 1993 the Queen's Awards for Environmental Achievement were conferred for the first time. Most of the twelve winning projects are of a "cleaner technology" nature. The RSA have, however, continued with an Environmental Management Award Scheme which I'm delighted to say my Department jointly sponsors. The first RSA Environmental Management Awards were presented by the Duke of Edinburgh a few days ago. It is significant to see that two out of the five RSA Awards went to companies that had developed policies of insisting that their suppliers provided detailed environmental information. The companies were B&Q and British Telecom. This sort of activity will underline the concept of a chain of responsibility from cradle to grave. It is also encouraging that two of the other awards and a commendation went to smaller companies. If environmental management in industry is to have a significant impact then it is vital to gain the commitment of the smaller business community. Award Schemes such as these help to keep environmental technology and management in the public eye and emphasise the importance which the Government attaches to these issues.

Conclusion: Sustainable Development

In conclusion, I should like to congratulate the RSC most warmly for organising this symposium. I can see from the programme that they have succeeded in bringing together an impressive list of speakers from manufacturing industry, academia and the consultancy sector. The symposium is particularly timely. In 1992 the Earth Summit put firmly on the agenda yet another key phrase: sustainable development. The UK is committed to producing later in 1993 a sustainable development plan as part of a worldwide United Nations initiative. Waste minimisation and cleaner technology are essential to achieving sustainable development. At the global level, failure would call into question the survival of human civilisation. The achievement of sustainable development depends on the skill and inventiveness of industrial scientists and managers, and the chemist has a crucial contribution to make. I hope this does not sound like an evasion of the responsibility of Governments. We in Government must certainly help by creating the right conditions in which industry can bring to bear its drive and creativity to deliver sustainable development. But we do all ultimately look to industry to deliver the ingredients for environmentally sustainable economic growth. Sustainable economic growth relates to an improving quality of life, including job satisfaction in the workplace. It <u>cannot</u> involve personal - or national - success being measured by conspicuous <u>squandering</u> of the Earth's resources.

I extend to everyone here my very best wishes for a successful Symposium. My sincere hope is that you take away from here a renewed enthusiasm for ever more efficient processes, based on the skills that are to be found in your vital sector of industry.

Waste Minimisation in the Context of Integrated Pollution Control

David Slater

DIRECTOR AND CHIEF INSPECTOR, HER MAJESTY'S INSPECTORATE
OF POLLUTION, ROMNEY HOUSE, 43 MARSHAM STREET,
LONDON SW1P 3PY, UK

The title of this talk is - "Waste Minimisation in
the context of Integrated Pollution Control" - so to
put the context "in context" I will start by saying a
little about Integrated Pollution Control (or IPC).

Integrated Pollution Control. In their Fifth
Report, the Royal Commission on Environmental Pollution
recommended that releases of pollutants from industrial
processes should no longer be subject to different
control regimes for different media. In their Twelfth
Report, they introduced the concept of Best Practicable
Environmental Option (or BPEO) which is a systematic
procedure for establishing which option, at acceptable
cost, is the most beneficial or the least damaging to
the environment as a whole, (in the long term as well
as the short term). As most of you will be aware, both
of these concepts were incorporated in legislation as
Part I of the Environmental Protection Act, 1990, ("the
Act" or EPA 90). This led to a rolling programme
starting with Large Combustion Plant in 1991, in which
industrial processes were to be brought under
Integrated Pollution Control for the protection of the
environment as a whole, - with HMIP as the prime
regulator.

In addition to introducing the principle of
holistic environmental protection and the associated
concept of Best Practicable Environmental Option, the
IPC part of the Environmental Protection Act also
introduced the fundamental principle of "Pollution
Prevention rather than Cure", and the right of the
public to have access to information about pollution
from prescribed processes. Important as the last point
is, it will be obvious to all of you here that we will
be mainly concerned today with the first two
principles, as these are likely to have much in common
with the principles behind "waste minimisation".

How are these IPC principles to be put into

practice? Part I of EPA90 sets out a series of
objectives which must be met in the operation of
prescribed processes. (Prescribed Processes, as most
of you will know, are those industrial processes
defined in Regulations as having the potential to cause
serious pollution, and therefore are to be subject to
Integrated Pollution Control). The first objective of
IPC, as Lord Strathclyde has just reminded us, is to
use the best available techniques not entailing
excessive cost (BATNEEC) for preventing releases, or
where that is not practicable, for minimising and
rendering the releases harmless.

The second objective aligns with IPC principle of
protection of the whole environment, ie. to ensure that
BATNEEC is used to minimise the pollution which may be
caused to the environment as a whole - the best
practicable environmental option (BPEO).

To meet these IPC objectives, HMIP's thinking and
approach to pollution prevention and control has
undergone a step change. This is illustrated by the
shift of emphasis from the Best Practicable Means
approach embodied in the Health and Safety at Work Act
(which was largely based on end-of-pipe abatement
technology), to the notion of using intrinsically
low-waste-producing techniques for the whole process so
that the harmful releases are minimised at source.

BATNEEC. So what does BATNEEC mean? I would
expect that most of you are, by now, familiar with this
acronym, - even if not feeling entirely confident about
its precise meaning. Starting with BAT part:

"Best" refers purely to the effect of the
technique on the environment, ie. it is that technique
(or those techniques, for there may be more than one
set of techniques with comparable effectiveness) that
is most effective in preventing, minimising or
rendering harmless polluting releases.

"Available" means procurable by the operator in
question. This doesn't imply that the technique has to
be in general use in the UK or overseas, but it does
recognise that novel techniques proven only at pilot
plant scale require a degree of business confidence
before they can be said to be available.

"Techniques" is defined in the Act. It embraces
not only the plant and equipment in the process, but
also the way it is operated, the manning and
supervision levels, the training and qualifications of
the staff etc.

So, how does HMIP come to a view on what constitutes
BAT? EPA 90 lays a specific duty on the Chief

Inspector "to follow developments in technology and techniques for preventing or reducing pollution of the environment due to releases of substances from prescribed processes". To this end, HMIP commissions research into available techniques either for specific production processes in which it has a particular interest or for wider industrial sectors. During the past year or 18-months, some of you may have been asked by consultancies to assist them with their surveys of techniques, current or being developed, for Chemical industry processes that are in operation in the UK. Similar reviews have also been carried out for Incineration, Mineral, Metallurgical and other prescribed processes - all for the purpose of providing input to the series of Chief Inspector's Guidance Notes to Inspectors.

These Guidance Notes, which are progressively being issued, give general guidance to HMIP inspectors on what might constitute BAT in those industries currently approaching authorisation under IPC. I must stress however that HMIP does not seek to prescribe the techniques which an individual operator shall use to achieve the Act's objectives. It is for the applicant to make his own assessment in relation to his own plant, its geographical location, etc and then to put forward his case for consideration by HMIP.

The Guidance Notes are not prescriptive; they merely represent a selection of possible alternatives which were available at the time of publication. If other techniques have come along since, or alternatives appear better in relation to a particular process then, of course, it is for the applicant to put them forward and demonstrate why they are the best available techniques.

So far I have spoken only about BAT. The NEEC or "Not Entailing Excessive Cost" part requires consideration of the cost of applying the Best Available Techniques in relation both to the nature of the industry concerned and to the degree of the environmental damage, and thus the cost of BAT that can be required before costs are considered excessive.

How does HMIP come to a view on what constitutes BATNEEC? For new processes BATNEEC will often be synonymous with <u>the</u> Best Available Technique, but for existing processes it is usually more complicated. There is an inherent expectation that all plant will, in time, be brought up to new plant standards (or ultimately closed down), so in addition to assessing the cost implications, my inspectors and I will be establishing time-scales for up-grading. The of progress may vary between one plant and another, depending upon the impact of their activities upon the environment.

   IPC Applications. As I have said before, it is
for the operator, or potential operator, of a
prescribed process to demonstrate in the application
for authorisation, that the chosen process techniques
meet both the BPEO and BATNEEC objectives. HMIP will
then make its own assessment, and attach conditions
before issuing (or refusing) the authorisation. The
Environmental Protection (Applications, Appeals and
Registers) Regulations, 1991 - SI 507 (1991), lay down
the minimum information that must be included in an
application before it can be accepted. Included in the
technical requirements are:

-   a description of the prescribed process,
-   a list of prescribed and other
    environmentally harmful materials involved in
    the process,
-   a description of the techniques used to
    prevent, minimise or render harmless any
    releases, and
-   (last, but definitely not least) details of
    any proposed releases, together with their
    environmental consequences.

   The official guidance note sent to all applicants
seeking authorisation, expands on the level of detail
required by HMIP to determine applications. Among the
items requested under the "process description" heading
are flow diagrams, reaction and side-reaction details,
mass balances of all materials in the process, and the
sources and magnitude of all releases. In other words,
water and releases of solid or liquid waste are
pre-requisites for determination of an IPC application.
Data also has to be provided for the range of operating
conditions - and for alternative options that may have
been considered.

   In that same guidance, reference is also made to
considering "improvements to the process operation, eg.
by changing from a raw material that may be
contaminated with a particularly harmful substance." -
an overt example of one of the maxims of Waste
Minimisation and pollution control - "If it ain't
there, you can't lose it!"

   These examples of HMIP's requirements for
quantification of releases and comparison of options,
taken against a background that insists on applicants
demonstrating that the BATNEEC and BPEO objectives have
been met, should indicate to you that we are not far
from a more formal "waste minimisation" assessment in
applications.

   Waste Minimisation in IPC Applications. I have

indicated that the operators have always had to include quantified assessments of all their releases, in their applications. However, in the most recent issues of Chief Inspector's Guidance Notes - ie. those covering the Chemical Industry - Inspectors are also advised to encourage applicants to carry out a formal process assessment and have in place a Waste Minimisation programme, in advance of submitting their application. In addition to identifying in a systematic way those areas in their process where reductions in releases may be accomplished, and thereby provide the foundations for a programme for up-grading the plant, the procedure provides a mechanism whereby applicants can identify deficiencies in the data to be included in their application (which is a problem HMIP has encountered far too frequently).

Waste Minimisation Guide. The introduction of the waste minimisation concept into those Chief Inspector's Guidance Notes that were issued after the I.Chem.E's "Waste Minimisation Guide" [1] was published in 1992, is no coincidence. The appearance of the Guide provided a mechanism (with defined methodology) by which HMIP could encourage operators of IPC processes to take the next step towards the basic principles of IPC, - namely "Prevention rather than cure".

As Lord Strathclyde mentioned, production of the I.Chem.E Guide was supported by the Department of the Environment. It was developed for the UK market by Prof Barry Crittenden's team from pioneering US EPA publications - the Waste Minimisation Opportunity Assessment Manual [2] and the draft Guide for an Effective Pollution Prevention Program [3]. It gives guidance on the practical techniques that can be implemented to reduce waste and provides the methodology that can be followed to drive waste minimisation programmes through to implementation.

I am sure that you will hear more about the methodology and the application of the Guide to Waste Minimisation programmes from later speakers. I will just comment briefly on key aspects that reflect HMIP's interest in the subject through IPC, and finish off with a few examples we in HMIP have encountered.

Waste Minimisation and IPC. What is waste? In its widest Waste Minimisation context, waste can be interpreted as almost any loss or discharge of any material to any medium. Under IPC, strictly it is only releases of substances prescribed in the Regulations which must be prevented or minimised; other releases need only be rendered harmless. However, the range of prescribed substances, particularly those prescribed for air, is large, so even if Release Minimisation is restricted to these areas, very significant

improvements can be made - and, once the thinking has
started, it is unusual for the programme to be
restricted solely to prescribed substances.

The concept of "Waste Minimisation", like IPC, is
based on the premise that Prevention is better than
Cure. Its main implications are:

-    The avoidance of reliance on end-of-pipe
     treatment of waste streams,
-    The adoption of "clean", low-waste
     technologies,
-    (and, where some waste production is
     unavoidable) Recycling, reclamation or use as
     a fuel.

What's involved in Waste Minimisation in practice?
Well, generally, waste Minimisation is <u>not</u> about
reducing waste after it has been generated, so the
following techniques usually have no place in a Waste
Minimisation programme:

-    Incineration,
-    Chemical transformation to a less harmful
     waste,
-    Biological treatment,
-    Transfer from one environmental medium to
     another where it might be less harmful,
-    Dilution or dispersion.

These techniques are end-of-pipe abatement options
which in a waste Minimisation programme are only
considered as measures of last resort to further treat
releases that have already been ministered.

So, what is waste Minimisation about? It is about
methodically assessing the process in question, using
the I.Chem.E/EPA methodology, with the aim of advancing
each waste - or release - producing part up the order
of priority from 6 to 1 in the following list of waste
reducing practices.

1   Elimination of waste production.
2   Reduced production of waste at source.
3   Recycling within the process.
4   Reclamation for recycling or other use.
5   Use of the waste for energy production.
6   Treatment to form less harmful waste.

<u>Waste Minimisation in practice.</u> What examples of
Waste Minimisation have the Inspectorate been involved
with? Lord Strathclyde mentioned the Aire and Calder
programme and whilst few, if any, of the improvements
have been directly connected with IPC applications,
(- it largely preceded the IPC timetable, and involved
many non-prescribed processes) I am very pleased that

HMIP has been involved in this "awareness raising" programme.

This is a demonstration project initiated by the Centre for Exploitation of Science and Technology (CEST) in March 1992 with financial support from Yorkshire Water, the BOC Foundation, and the NRA as well as HMIP. Eleven companies in the Aire and Calder catchment area are actively involved, with the assistance of March Consulting, in waste reduction programmes. So far, over £2m worth of environmental and resource savings have been identified, mainly from reduced demand for water, energy and effluent treatment. Over half of the savings have arisen at a relatively new site which was previously considered to be both efficient in its use of resources and environmentally friendly in respect of waste produced. However, systematic waste minimisation assessment has completely removed some water-using and waste-producing activities ("Waste elimination") and changed other waste-producing activities by "Reduced waste production at source", "Recycling" or "Reclamation". Other companies have used practices 1,2 and 3 to make dramatic improvements in their discharges to sewer or controlled waters - and much improvement has come from simple, virtually zero-cost, "housekeeping" procedures.

Many of the companies participating in the Aire and Calder programme have discovered what waste Minimisation can achieve:

- Environmental protection <u>and</u> Reduced Production costs.
- Less Pollution <u>and</u> Less Resource Consumption, - by reducing waste at source or by recycling it.

In other words they have found "Win/Win" scenarios.

In the case of companies that have made applications for IPC authorisation, a striking example of one where a "dirty" process has been replaced by "clean" technology is that of a petrochemical company in Southern England. The company required pure butenes for its down-stream processes but the butene supply was heavily contaminated with butadiene. They had a requirement for butadiene so operated a wet extractive process using cuprous ammonium acetate to separate the butadiene from the butenes. This process typically released annually around 200t of ammonia and 140t of VOCS to air, and around 300t of ammonia and 6t of copper to the aqueous effluent treatment system. Involving HMIP at the design stage, they resolved to source their butadiene from elsewhere and replace the

butene purification process with a new one having
essentially zero emissions. Authorisation has been
given for a catalytic hydrogenation process which
selectively reduces the butadiene to the required
end-product butene. There are no emissions to air
other than those estimated by the standard SOCMI method
for fugitive releases from flanges, valves, pump glands
etc, (around 7tpa VOC) and they estimated that less
than 1kg per year of hydrocarbons is released to the
water course.

Another example of a company changing the process
completely to eliminate the release of a prescribed
substance - in this case VOCS - is that of an
American-owned manufacturer of fluorescent tubes. The
company is currently preparing an application for
authorisation as a section 4.5 "Mercury process". The
insides of fluorescent tubes are coated with a phoshor
that has traditionally been applied as a solution or
suspension in a xylene/butanol based solvent.
Following the drying/curing process, around 500t per
year of solvent is lost to atmosphere. Having come
under near-simultaneous environmental pressures on both
sides of the Atlantic, the laboratories of the parent
company put around 30 man-years of effort into finding
a "cleaner" process. They have now developed a
water-based carrier system for phosphor deposition
which is to be installed in the UK factory. The only
significant release expected is 1.25t per year of
ammonia loss to atmosphere. The capital cost of this
change of technology is high so the company is
replacing its four existing production units in a 4
year rolling programme.

More typical of HMIP's experience to date are
cases where, at pre-application discussions with
operators, it has become clear that releases of some
materials - particularly VOCS to air - have never been
considered particularly important. When the companies
concerned have quantified the releases (and often been
surprised by the size) and have been pointed in the
direction of Waste Minimisation by Inspectors, they
have often come up with solutions which do not cost
much to implement or are even "Win/Win". They use less
organic solvent in the first place or they are able to
recover at least the fuel value of the VOCS that were
previously lost.

Conclusion. I will conclude by reminding you that
Waste Minimisation is a "Motherhood" type of concept;
nobody is against it, but it is still remarkably
difficult for some people to adapt their ways of
thinking to make it the "norm". HMIP made a
step-change in its thinking when it started to regulate
within the IPC framework, and the same sort of culture
change that the Waste Minimisation concept requires,

has to spread right across industry. Much progress is being made - as I am sure we will hear from other speakers today and tomorrow.

Inspectors will continue to use IPC applications (as well as any other opportunities that arise) to encourage operators to carry out systematic Waste Minimisation assessments because, as the Aire and Calder programme and others have shown, it is often possible to reduce costs at the same time as improving environmental protection. It is indeed often possible to find a "Win/Win" scenario.

References:

1.   Waste Minimisation Guide. 1992. The Institution of Chemical Engineers, 165-171 Railway Terrace, Rugby, CV21 3HQ.

2.   Waste Minimisation Opportunity Assessment Manual. 1988. US EPA Hazardous Waste Engineering Laboratory, Office of Research and Development, Cincinnati, Ohio.

3.   Draft Guide for an Effective Pollution Prevention Program. 1991. US EPA (at the address for reference 2).

Waste Minimisation and Clean Technology – Why Do It?   A CIA View

M. Wright

IMPERIAL CHEMICAL INDUSTRIES PLC, GROUP HEADQUARTERS,
9 MILLBANK, LONDON SW1P 3JF, UK

1   INTRODUCTION

The public has accepted the contribution that the products of the modern chemical industry make to health, shelter, food, mobility and the reduction of drudgery. There is also recognition of its position as premier exporter (£15bn in 1992) and major employer (300,000 jobs). However, public concern about production, use and disposal of chemicals flares up rapidly whenever there is an incident and the holistic approach to husbandry of the earth's resources and avoidance of waste has acquired firm foundations with the young. The environmental movement has used public concern to press successfully for legislation to regulate potentially polluting activities.

Many of these issues are highly complex which has invariably produced arcane legislation quite beyond the comprehension of non-experts and invariably adding to the burdens of industry. Nonetheless it is vital for the chemical industry to be seen to be well regulated and this paper addresses some of the reasons why we must commit to improved environmental performance, some examples of successful programmes and a few pointers to where we may be headed longer term.

2   PUBLIC PERCEPTION OF THE CHEMICAL INDUSTRY

The latest view of the public, through opinion polls (MORI Autumn 1992)[1] puts economic factors such as jobs and prices top of the agenda but concern for the environment has shown remarkable resilience during recession and is regarded as the second most important area of concern that government should be paying attention to and sixth most important for industry. The importance of the environment is a key feature in the public's favourability rating of the chemical industry which has fallen from 50% in 1979 to 21% in 1992 although the corresponding unfavourability rating seems now to have stopped rising.

The vast majority (83%) believe that use of chemicals causes damage to the environment but half have the confidence that industry is working hard to solve the problem. It's worth noting that most people receive their information from TV documentaries and news but after these sources the highest credibility rating is given to information from family or friends who work in the industry.

The chemical industry has to regard environmental protection as a business imperative, and indeed, the legislators, particularly in Brussels, will continue with a heavy programme of instruments in response to public pressure.

3   GOVERNMENT RESPONSES

EC Fifth Environmental Action Programme[2]

Within the EC the community has had four environmental action programmes from 1977 up to end of 1992. However, with its fifth action programme, instead of trying to solve pollution problems through concentration on environmental controls, it aims to change the human behaviour which cause these problems by adopting a set of long-term quality objectives. Measurable targets are to be set up to the year 2000 and the 'principal actors' are called upon to implement actions to achieve the targets. The principal change areas are industry, energy, transport, agriculture and tourism and a broad range of instruments is envisaged ranging from legislation through market measures, financial support and education programmes. The overall objective, enshrined in Maastricht, is to promote 'sustainable and non-inflationary growth respecting the environment' and environmental policy "must be integrated into the definition and implementation of other community policies".

Dutch National Environmental Policy Plan[3]

A good illustration of this approach is the Dutch Government which had been collecting emissions data since 1985 and was thus able to produce a quantitative National Environmental Policy Plan (NEPP) in 1989 which included emission reduction targets together with estimated costs for their achievement. The NEPP is now being implemented in Holland through a series of voluntary agreements with different industry sector groups. The issues included in NEPP are climate change, acidification, eutrophication, emissions to air, water and soil, disposal of waste and disturbance such as noise and odour. The Dutch Chemical Industry Association (VNCI) together with individual chemical companies recently signed a voluntary agreement with the Government committing to achievement of detailed targets by the year 2000, covering inter alia (see charts 1-4) Climate Change, Diffusion-Air, Diffusion-Water, Waste

CHART 1

Climate change

Reduction targets (perc.)

- ▸ CFC's, halons   100   (1995)
- ▸ Carbon dioxide*   3-5   (2000)
    Energy efficiency*   20   (2000)

* 1989 / 1990 = 100

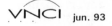 jun. 93

CHART 2

Diffusion - Air

Reduction targets* (perc.)

	2000	Intention 2010
Ethylene	50	90
Acrylonitrile	50	97
Benzene	75	97,5
Toluene	50	90
Monovinylchloride	90	90
Dust	75	95

*1985 = 100

VNCI jun. 93

CHART 3

Diffusion - Water

Reduction targets * (perc.)

	1995	2000	Intention 2010
Acrylonitrile	-	50	90
Benzene	60	75	90
Oil	60	60	90
Zinc	50	65	80
Nickel	50	50	80
Lead	65	65	70

* 1985 = 100

 jun. 93

CHART 4

Waste disposal

Targets (perc.)	1995	2000
Prevention)		
Recycling)	40	90
Re-use)		
Landfill	50	7
Incineration	10	3

VNCI jun. 93

Disposal. Potential advantages are seen to be the
certainty of the targets which allows planning of
process-integrated 'clean technology' and gives time for
development of cost-effective solutions integrated across
air/water/waste and implemented over a sensible timescale.
A steering group is set up between government and industry
to ensure fairness and consistency across the country.

UK Statutory Water Quality Objectives

In the UK the government is using a consultative approac
for example the development of Statutory Water Quality
Objectives (SWQOs), in the first instance for rivers. In
1992 the NRA developed a proposal for controlled waters
which was adopted by Government. The DoE are currently
drafting regulations for the classifications and will then
be testing these in some pilot river schemes. The process
will involve consultation with individual dischargers who
will have to meet tighter standards and cost-benefit
analysis will figure prominently in deciding on proposed
action.

4   EXAMPLES AND LESSONS

Chlorinated Hydrocarbons Emissions Reductions - Merseyside

However, all human activity creates pollution and
implementation of programmes to reduce emissions will
often generate side effects which will need to be measured
and balanced against the nature of the original problem.
I can illustrate this with a current project at ICI in
Merseyside which is designed to reduce emissions of
chlorinated hydrocarbons (CHCs) to the atmosphere, to the
River Weaver, and to enable closure of a lagooning
operation. After securing maximum reduction at source ICI
engineers reviewed all the available technologies and
selected a combination of air stripping to remove CHCs
from aqueous effluents and incineration to remove CHCs
from plant vents and air stripping vents. The advantages
were that the technologies enabled maximum recycling of
chlorine, maximum energy recovery and very low energy
usage to drive the technology. The net effect of the
project will be to reduce discharges of CHCs by 10,500
tonnes/annum to air and 27 tonnes/annum to water at a cost
of around £40 million. Local environmental groups are
approaching the local community through the distribution
of leaflets etc, citing the dangers of dioxins to oppose
incineration. However, at the end of the day the public
must decide. We believe that the local community
understand that we are taking a responsible position and
that we are looking to improvement in amenity which will
bring real benefit. Any major project of this nature must
expect to devote major resources to explaining its case to
the public authorities and to the local community.

Sulphuric Acid Recycle - Billingham

A second example where there has been much debate about the environmental effects of the solution to a problem concerned the disposal of a large quantity of dilute acidic ammonium sulphate waste from a process which leads to production of the key upstream chemical for production of 'Perspex' acrylic sheet manufacture in NW England.

After lengthy scientific studies in conjunction with MAFF it was decided that this waste could be sent to sea disposal. The sea has enormous buffering capacity, animals excrete ammonia and the sulphate ion is present in large quantities. Extensive studies of the fish, plankton and benthic (bottom dwelling) creatures in the disposal area showed they were more healthy than a control sample from clean waters off the SW Coast of England and we argued this disposal was Best Practical Environmental Option (BPEO). However, the perception of potential harm, fostered by the environmental movement and some of the other North Sea states, eventually led to the decision of the 1990 Ministerial Conference on the North Sea in the Hague to cease sea disposal by the end of 1992. As a result, we have just commissioned a £66m sulphuric acid recycle unit at Billingham which has enabled us to cease sea disposal of this waste. The new plant is a net user of about 40,000 te/yr of natural gas with commensurate Carbon Dioxide emissions. There were many other business development projects competing for these resources.

5   COSTS AND COST-EFFECTIVENESS

UK Chemical Industry Spending on Environment

Recession has taken its toll on fixed capital expenditure by the Chemical Industry in the UK with a fall of about 11% in 1992 on 1991 levels. However, the proportion attributed to environmental protection spending has increased from 9% in 1990 to 14%. (See chart 5).

Environmental operating costs have also increased and combined capital and revenue costs were around 3.5% of total turnover in 1992 (all CIA data)[4,5].

Search for Greater Cost-Effectiveness

These economic realities have put scientists and engineers under pressure to devise innovative, cost-effective solutions to their environmental problems. This is a fast moving area and at ICI we have needed to make use of the many-fold times more work going on externally around the world in universities, government laboratories and with other companies who were willing to collaborate with generic, pre-competitive work. To achieve this we have established an environmental

CHART 5

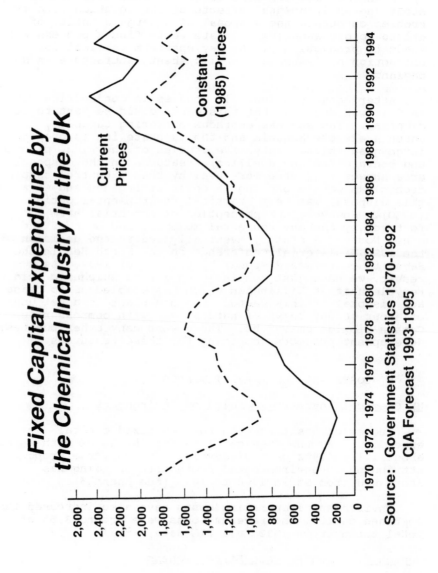

Fixed Capital Expenditure by the Chemical Industry in the UK

Current Prices

Constant (1985) Prices

Source: Government Statistics 1970-1992
CIA Forecast 1993-1995

technology brokerage system through individuals charged
with linking solutions to problems and with looking for
opportunities for new business. In the USA we have
participated in 'Project Listen' which is a collaboration
system for these environmental technologies. Our
engineers are managing 'Project IMS' (Intelligent
Manufacturing Systems) which is a series of international
collaborations involving European, N American and Japanese
companies sharing eg reviews of regulations and best
environmental technologies to satisfy needs, eg for
incineration.

Each of these schemes is aimed at achieving maximum
environmental improvement for every £ spent. At the end
of the day the public will have to decide, probably
through Parliament in our case, what standard of
environmental quality is needed and what rate of progress
can be afforded.

6   ENVIRONMENTAL PRESSURES AND RESPONSES

Three key themes used by environmentalists to press for
environmental improvement have been:

·   The Precautionary Principle: which at its most extreme
     requires total containment of waste with zero
     emissions and at its most reasonable requires detailed
     assessment of each emission to the environment to seek
     reassurance that no harm will be caused to man or the
     environment.

·   The Proximity Principle: which presses for solutions
     to the disposal of waste at the point of their
     creation to avoid the hazards associated with movement
     and, in particular, to avoid wastes being moved to
     inadequate places of disposal.

·   The Principle of Openness: those who make emissions
     deemed potentially harmful to the environment without
     quantifying and declaring them are attacked as
     uncaring and, potentially, incompetent.

These themes were picked up in 1990 by the Chemical
Industries Association in its booklet 'Chemicals in a
Green World' which responded with voluntary action with
the 'Responsible Care' programme as its centre-piece.
Other contributions were included on government policy,
instruments, waste management strategy, greenhouse gases,
protecting the aquatic environment etc, which helped shape
the UK government White Paper "Our Common Future" and the
EPA 1990 with Integrated Pollution Control as its
centre-piece. At EC level the proactive stance of the
chemical industry has enabled a beneficial contribution to
development of the recent Eco management and audit
regulation and the industry's views have had a significant
effect on proposed Directives on Integrated Pollution

Prevention and Control (IPPC) and Polluting Emissions
Registers (PER). Industry has expressed its strong desire
to work with government on proposed regulation and
deregulation.

7   VOLUNTARY ACTION PROGRAMMES

CIA 'Responsible Care'

Having worked hard to influence legislators to produce
workable regulation we must, of course, comply with the
letter and the spirit of the law.   In addition, in the UK
chemical industry the CIA through its 'Responsible Care'
programme has made a voluntary commitment to demonstrate
continuous improvement in all aspects of health, safety
and environmental performance.

The CIA guiding principles formed the basis of this
commitment; ie companies should:

-    assess the actual and potential IMPACT of
     activities and products.

-    work closely with public and statutory bodies to
     achieve an acceptably high level of health, safety
     and environmental protection.

-    make available relevant information.

Indicators of Performance

Any quality system needs to measure, manage and
improve and today in London, CIA is presenting its first
set of indicators of performance against its 'Responsible
Care' commitments[5].   The environment indicators cover the
period 1990-1992 and include the spending information
which I have shown; data on discharges of red list
substances to water which decreased by 40%; disposals of
special waste to landfill which decreased by 9%; and
energy consumption indices which demonstrate a 60%
improvement over 20 years.   Short term energy consumption
per unit of output levelled off last year as a result of
reduced volume caused by recession.

US EPA 33/50 Programme[6]

On the other side of the Atlantic the US Environmental
Protection Agency has developed a voluntary pollution
prevention programme called 33/50 which has targeted 17
chemicals (see table) chosen on the criteria that they are
high volume industrial chemicals which pose some
environmental and health concerns and their emissions can
be reduced through pollution prevention.

Table 1   US EPA 33/50 Programme Target Chemicals

Benzene	Methyl Ethyl Ketone
Cadmium and Compounds	Methyl Isobutyl Ketone
Carbon Tetrachloride	Methylene Chloride
Chloroform	Nickel and Compounds
Chromium and Compounds	Tetrachloroethylene
Cyanides	Toluene
Lead and Compounds	Trichloroethane
Mercury and Compounds	Trichloroethylene
	Xylenes

EPA is asking companies to commit voluntarily to reducing emissions of these chemicals (total to air, water and land) by 50% in 1995 with an interim goal of 33% in 1992. By end 1992, more than half of the USA's largest releasers of 33/50 chemicals had sent commitments in writing to EPA and were working to make the reductions a reality.

The advantages of this voluntary programme for industry were seen to be:

- creation of clear national goal reductions for targeted chemicals.

- flexibility to choose cost-effective environmental solutions which may result in improved efficiency and net economic benefits.

The advantages for the public were faster reductions of emissions than might otherwise occur and promotion of a pollution prevention ethic in American business.

8   ENVIRONMENTAL REPORTING

CEFIC Draft Guidelines[7]

The European Chemical Industry Council (CEFIC) has responded to the need to provide the public with environmental performance information by drafting guidelines for its member companies indicating the families of substances whose emissions to air, water and to land should be reported upon.

Table 2

RELEASES TO AIR	RELEASES TO WATER	WASTE TO LAND
Particulates	Suspended Solids	Landfill
SO_2	TOC	Other Disposal
NO_x	N	Non Hazardous
VOCs	P	Waste
Heavy Metals	Heavy Metals	
	EC List 1 Substances	

Guidance has also been drafted on the proposed sections which would be covered in environmental reports, eg Corporate Reports would cover items listed on the table:

Table 3   Corporate Reports

Company scene setting policy etc	Environmental management – systems and organisation
Production development	Emission data
	Energy data
Product developments	Health and Safety data
	Communication processes
Plans, objectives, goals	

Environmental performance reporting for individual sites of a company is designed to meet a different set of needs, ie it is focused on the local community and would cover:

Table 4   Site Reports

Site scene setting - policy etc	Emission data
Site set in local perspective eg economics, services, local environmental features	Energy data
	Health and Safety data
Environmental management systems	Complaints
Organisation, objectives, protective techniques, emergency plan	Communications, open days etc.

9   MORE FROM LESS

Improved In-House Technology

For decades the chemical industry has engaged in in-process recycling and re-use of materials and energy for example the amount of energy needed to make Ammonia has dropped from 88 GJ/te for coke-based technology to 28 GJ/te for recent ICI processes and emissions of NO_x to air reduced by approximately 90%.

Over the next two days you will be hearing many impressive examples of waste reduction achieved by, for example, mimicking natural systems which may be fuelled by solar energy and operating at ambient temperatures and pressures using novel catalyst systems, or from the results of vigorous 'shoe-leather' management such as in our Paints business where systematic measurement of all waste generated, combined with individual improvement projects, eg for VOC emissions, clean solvents, sludge to landfill and even used drums, bags and packages has reduced waste by 30% in 1992 versus 1990.

Post-Consumer Waste

Greater prominence in the public eye comes from post-consumer waste recycling and this has achieved headline status in countries where unilateral regulatory action has been taken, for example the German Packaging waste laws. Levies are imposed to ensure that packaging waste is disposed of by manufacturers to the extent that these recycled materials are being exported in such large quantities that they are disrupting the waste recycling industries in neighbouring countries, eg France and UK.

Life Cycle Assessment

Much thought is being given to quantifying environmental effects of products through the development of Life Cycle Assessment methodology and major progress has been made towards agreeing inventories of effects, which currently are being used in Eco-labelling exercises, but there is still much work needed on risk assessment of the environmental impacts of activities.

A classic example is paper recycling where paper produced from managed, replanted forests, eg in Norway, which is converted in mills where the timber waste powers the process, should be burned when finished with by the consumer to raise energy in power plants. In contrast, paper should be recycled when it has been produced from ancient forests which are not being replanted. The equation depends upon where the boundaries are set during the assessment.

In its recent Seventeenth Report the Royal Commission on Environmental Pollution recommends recovery of energy from municipal waste by incineration as long as good standards of operation and control are observed.

10   OUTLOOK

Closed Loop Systems

Improved scientific understanding of environmental problems and their solutions will force the pace for more closed loop systems.

In processing for example, combined heat and power systems which are typically about 70-75% efficient at recovering energy from fuel versus 30-35% for a traditional power station will be linked with low grade heat projects, eg fish farms. Short life articles such as plastic bottles will be made from plastics which can be recycled; eg Polyethylene Terephthalate (PET) can now be collected, methanolysed, and re-polymerised economically back into bottles which have been approved by US Food and Drug Administration.

Opportunity for Competitive Advantage

This approach to 'industrial ecology' is consistent with the global goal of sustainable development and corporations which speedily understand and address these problems early will generate opportunity for competitive advantage.

11   CONCLUSION

In summary, pressure for improved environmental performance is here to stay and cost effective compliance with legislation will be a competitive necessity. In addition it will be crucial to plan to deal with existing problems over an economically feasible timescale and really perceptive operators will find new and rewarding business opportunities from their response to environmental issues. This will all be done in the public spotlight where there will be a presumption of openness of information and industry will need to establish effective dialogue with all affected parties.

REFERENCES

1   Research Study conducted for the Chemical Industries Association, Autumn 1992.

2   "Towards Sustainability", a European Community Programme of Policy and Action in relation to the Environment and Sustainable Development, COM(92)23 Final, Brussels, 27 March 1992.

3   "To Choose or to Lose" National Environmental Policy Plan 1990-1994, Ministry of Housing, Physical Planning and Environment, The Hague, The Netherlands.

4   CIA Investment Intentions Survey Results, 1993.

5   CIA, "The UK Indicators of Performance 1990-92".

6   "Forging an Alliance for Pollution Prevention", United States Environmental Protection Agency, August 1992.

7   European Chemical Industry Council - CEFIC Guidelines on Environmental Reporting for the European Chemical Industry - Draft May 1993.

Corporate Strategy to Waste Minimisation

Efforts of BASF towards Cleaner Technology and Waste Minimisation

K.-G. Malle

BASF AG, CARL BOSCH STRAßE 38, D-6700 LUDWIGSHAFEN, GERMANY

1 "END OF THE PIPE"-ENVIRONMENTAL PROTECTION

At a chemical complex like Ludwigshafen, employing 45,000 people in some 350 production plants, it inevitably becomes clear at an earlier stage than elsewhere that the protection of the environment calls for long-term solutions.

Even during the reconstruction of the works in the 1950s, plants were being fitted with the first filters for waste air purification. Today, about 2500 waste air purification units of all kinds make it safe to breathe deeply in the Ludwigshafen conurbation. A conspicuous example is the 300 MW-Power plant of BASF with a 200 Mill. DM desulfurization- and a 110 Mill. DM dinitrification-unit.

A substantial proportion of air pollution originates from the generation of energy. This makes it particularly important that energy should be used economically. At BASF AG only less than one third of the necessary 16.2 million tons of steam in 1992 had to be produced from fossil sources: coal, oil or gas, compared with still two thirds twenty years earlier. The rest was regenerated waste heat or derived from waste incineration.

As early as 1959, BASF commissioned the first investigation dealing with improving the waste water situation of the whole complex. A schedule was drawn up, envisaging the construction of a second effluent drainage network, 35 km in length, and a treatment plant which was

also designed to benefit surrounding towns and munici-
palities. The uncontaminated cooling water and production effluent were completely separated and subjected
to continuous monitoring. The program was completed on
time when this plant, the largest of its kind on the
Rhine river began operation at the end of 1974.

The first incinerator for solid residues was lit in
1960. In-depth research work resulted in the incineration technology which is now used worldwide for this
task - rotary kilns with afterburn chambers. The eighth
unit has now come on stream, increasing total capacity
at Ludwigshafen to 175,000 metric tons per year.

BASF began using its own landfill, on the island of
Flotzgrün in the Rhine, in 1966. The site was carefully
selected to allow the environment-friendly transporta-
tion of waste by barge. Conscientious pre-sorting and
meticulous landfill technology prevent the creation of
new pollution. Today, every additional landfill section
is sealed off from the subsoil by a double liner of
plastic sheeting. This allows any leachate to be collected and treated, while leaks should they ever occur
can be traced and sealed.

The BASF subsidiary Kali und Salz AG has an underground
dump at Herfa-Neurode. This disposal facility for special wastes has become indispensable today, indispensable in fact for Western Europe as a whole. BASF itself
makes relatively little use of this dump: well under
1 % of the solid production residues from Ludwigshafen
are stored there.

No waste is brought to other dumps or even shipped to
foreign countries.

If we consider specific environmental pollution, these
initiatives have successfully broken the link between
production and environmental pollution at Ludwigshafen.
While the volume of production has risen by more than
40% since 1972, waste water pollution has fallen over
the same period by 93%, and atmospheric emissions by
75%. The start-up of the effluent treatment plant in
1975 meant that the proportion of solid residues taken
to landfill initially increased sharply, because of the

production of sewage sludge, but it has since fallen by more than 60%.

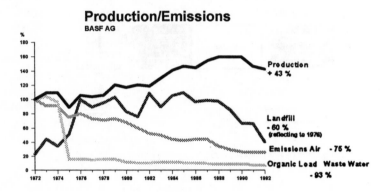

Figure 1 Production/Emissions

Looking back we have to appreciate that this "end of the pipe"-philosophy resulted in a rapid direct improvement of the situation. In the future, too, this will remain the key task. It has to be tackled jointly by the consumer, who generates the residue, as well as by the manufacturer of the products, who possesses special knowledge of their properties, and by specialist disposal companies. Each individual must be made aware that a high living standard and high levels of consumption inevitably require the provision of disposal facilities. Residues cannot be talked away or made to disappear by regulations. Exporting waste to neighboring countries is no solution for a highly developed industrial state. Politicians and the public, too, must help to prevent the attitude of "not in my back yard" from becoming the guiding principle for action.

Environmental Protection
End of the Pipe

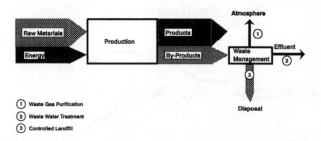

<u>Figure 2</u> Environmental Protection, End of the Pipe

2   ENVIRONMENTAL PROTECTION INTEGRATED INTO THE
     PRODUCTION PROCESS

On the other hand the "end of the pipe"-philosophy has
technical limitations. Each measure has a finite efficiency, in turn often involves additional pollution,
for example as a result of energy requirement, and
reaches the limits of economic good sense all the more
quickly, the higher the requirements to be met. Moreover, problems are sometimes only shifted, as when air
pollutants are washed into waste water, volatile substances from the water are emitted into the atmosphere,
etc.

To make further progress in environmental protection,
greater importance must inevitably be attached to a way
of thinking which embraces all media and in which the
quantity and disposal of the undesirable anthropogenic
substances are the key factors, "the best practical environmental option". Furthermore, the industrial process itself, which gives rise to the need for disposal,
must be included in the optimization considerations
from an environmental point of view. Sparing use of raw
materials, minimization and utilization of the by-products and maximum energy recycling are the objectives.
Environmental protection is integrated into the process
so as to embrace all media.

**Environmental Protection
Combined with the Production Process**

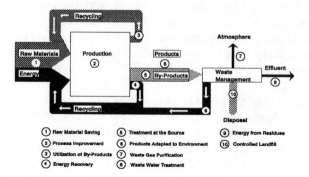

<u>**Figure 3**</u> Environmental Protection Combined with the
Production Process

Today, more than ever in the past, the central objective of research and development is to develop processes which result in the production of fewer residues, and processes which can make profitable use of recycled residues. The rapidly increasing costs of all disposal services are in themselves an effective incentive here from the market economy standpoint. The convenient formula for this today is: avoidance, reduction, recycling.

3 AVOIDING BY-PRODUCTS BY PROCESS IMPROVEMENT

Process modifications which lead to lower environmental pollution are the optimum solution to environmental problems. Residues which can be avoided do not have to be disposed of. Experience has shown that in large modern continuous plants even environmental protection problems can be minimized. In this respect, rapid innovation also means effective environmental protection.

In general, the new developments relate only to steps in a production process, the significance of which is difficult to describe to the outsider. Frequently, the development is subject to company secrecy. For these reasons, not many of the improvements constantly taking place everywhere are being published. Here are some examples.
Hydroxylamine, one of the intermediates in nylon production, was previously synthesized from sulfite and

nitrite by a process developed by the chemist Raschig. In the production of hydroxylamine, the waste water contained neutral salts, the amount of which was several times the amount of hydroxylamine. On the other hand, in the catalytic hydrogenation of nitric oxide in sulfuric acid, only small amounts of neutral salts are formed as by-products.

Methacrylic acid is the starting material for acrylic glass, which is made by polymerisation of the methyl ester. The acid is normally produced from acetone and hydrogen cyanide. A cyanhydrine is formed as the intermediate. When this is hydrolysed to give methacrylic acid, the by-product ammonium sulfate is obtained in larger quantity than the desired acid.

The BASF process makes use of two processes which in Ludwigshafen also serve for other purposes, namely hydroformylation and the addition reaction of formaldehyde. Methacrolein is obtained, which can easily be oxidized to methacrylic acid, avoiding any ammonium sulfate by-product.

Tert.-Butylamine, a starting material in the production of rubber chemicals and pesticides, is normally produced from isobutylene by reaction with hydrogen cyanide with the formamide as intermediate, the so called Ritter-Process. Equal amounts of sodium formate by-product are obtained.

BASF developed a catalytic process in which the tert.-Butylamine is produced by direct reaction of isobutylene with ammonia in an excellent yield.

High purity paraffins have a number of uses in the pharmaceutical and food industry. Hitherto the raw paraffins were purified in a complicated multistage procedure using bleaching clay, which has to be landfilled after use, and sulfuric acid which was subsequently incinerated.

The new process consists of a one step catalytic hydrogenation with a nickel-molybdenium catalyst on a special aluminium oxide carrier. It was possible to increase the yield by 10 percent with this purification process.

In fertilizer production phosphate rock has to be made
soluble, whereby the farmers prefer high phosphorus
and nitrogen contents in the product. Usually the rock
is in a first step processed with sulfuric acid and the
unsoluble gypsum formed discharged with the waste water
or landfilled. In a second step the phosphoric acid is
used for treating additional phosphate rock. The fer-
tilizers produced normally contain too little calcium
with the consequence that an additional calcium supply to
the fields is necessary.

The BASF process uses nitric acid for reaction with the
rock, avoiding the gypsum by-product and resulting in a
calcium-rich product.

4   RECOVERY AND UTILIZATION OF BY-PRODUCTS

Large chemical complexes have grown up throughout the
world because the integrated utilization of raw materials, by-products and energy has cost advantages. The
larger the complex, the better the prospects of being
able to use the by-products of one unit in some other
unit. Experience shows that there is a substantially
greater potential for reducing the amount of residues
by utilizing previously unused by-products than by improvements to the process. The utilization of by-products is particularly advantageous when the product is
required in a much greater amount than the amount of
by-product produced.

In the oxidation of o-xylene with air to give phthalic
acid, an intermediate for plasticizers, a large amount
of by-products is formed; these are removed from the
off-gases by scrubbing with water. In the past, this
concentrated waste water had to be burnt, using additional fuel. A process has now been developed by which
maleic acid is obtained as a by-product, which is
needed in large amounts in the production of plastics.
In addition the washing water can be recycled, so that
waste water no longer has to be purified.

In a similar process, cyclohexane is oxidized with air
to give an anol/anon mixture, an intermediate in the
production of nylon-6. This process also generates a
large amount of by-products. They too were extracted

with water and burnt. A process was developed to separate adipic acid from the waste water; this acid can be
used for nylon 66-production. In a second step the
mother liquor is oxidized with nitric acid so that a
mixture of succinic and glutaric acid is formed, which
can be isolated and used as a mixture in varnish production. The nitric acid can be destilled and recycled,
so that only a residue remains, which is incinerated,
with energy recovery.

Acetylene is produced from natural gas by a partial
oxidation process. The off-gas consists mainly of hydrogen and carbon monoxide. For a long time this gas
was only used for steam generation. Since 1988 more
than 80 % of the gas has been fed as raw material to a
methanol plant. The advantage compared to the direct
production of methanol from natural gas is that the energy for steam reforming is saved. Thus, with a capacity of 740 metric tons of methanol per day it has been
possible to save 3500 GJ energy and avoid emission of
about 720 kg of nitric oxides and 600 tons of carbon
dioxide.

When long-chain olefins react with hydrogen and carbon
monoxide at high pressure and temperature, the so-called
hydroformylation process, not only aldehydes which in a
second step can be hydrogenated to the corresponding
alcohols, are formed. In a side reaction between 3 -
15 % of the saturated hydrocarbons are produced. They
used to be burnt in the BASF power plants. Since then
BASF has succeeded in using this mixture as raw material in the steam cracker, saving the corresponding
amount of expensive naphtha.

The crude gas from the above-mentioned cyclohexane oxidation process still contains a mixture of acidic organic compounds and has to be purified with a sodium
hydroxide solution before further use. A process was
developed in which the resulting concentrated caustic
solution is burnt with energy recovery, and the sodium
carbonate which remains can be isolated and sold as
high grade material after it has been recrystallised.

In the production of synthesis gas ($CO + H_2$) from heavy
naphtha the waste water contains a large amount of soot

and ash. In the past it had to be filtered, and the filter cake was landfilled.

BASF found that the filter cake is an excellent filtration accelerator for the sludge of the sewage treatment plant, and that even the energy can be recovered by subsequent incineration of the sludge.

BASF operates a coal-fired power plant with a capacity of 800 MW. Normally such off-gas is desulfurized by scrubbing it with milk of lime, this would have resulted in 35 000 metric tons of gypsum annually.

We therefore decided to apply the Wellman-Lord process in which, in a first step, the sulfur dioxide is dissolved in a sulfite solution, forming bisulfite. In a second step the bisulfite is thermally cracked again to sulfite and highly concentrated sulfur dioxide, which can be liquified and reused as a raw material in several processes. A great deal of process development was and still is necessary to minimize the formation of sodium sulfate by-product.

5   UTILIZATION OF BY-PRODUCTS ALSO ENTAILS A RISK

However, the utilization of by-products means not only an opportunity but also a risk. An inescapable by-product creates technical as well as economic interdependence. What are the technical possibilities and what sort of price structure is needed when one product may be more in demand today and another product tomorrow? The alkali chloride electrolysis, in which equimolar amounts of chlorine and sodium hydroxide are obtained (in addition to hydrogen) is a textbook example of this. In the past, the demand situation with regard to chlorine and sodium hydroxide changed several times, both of these substances having many applications. The idea that chlorine is only a troublesome by-product of sodium hydroxide production is by no means true.

6   PREVENTION AND RECYCLING OF RESIDUES

During the last 15 years BASF succeeded in reducing the quantities of residues produced by 640 000 tons per year as a result of process improvements, by more than 1 Million tons as a result of physical recycling and by

175 000 tons by incineration and recovery of
the energy content.

Prevention and Recycling of Residues
last 15 years

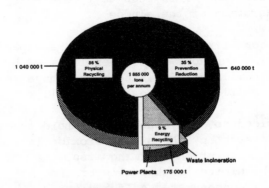

Figure 4 Prevention and Recycling of Residues

There is of course no end to such process optimization.
There will always be new developments which further reduce the environmental pollution. On the other hand we
should not be under any illusions: there are limits to
human ingenuity, and natural laws cannot be circum-
vented by cunning. In the future there will still be
solid and liquid residues to be disposed of.

7   INTEGRAL ENVIRONMENTAL PROTECTION

The greater the advances in environmental protection
during manufacture the more the focal point shifts to
environmental protection with regard to the use of the
products of the chemical industry. It becomes obvious
that the products themselves can contribute to environmental pollution during their further processing, their
use and finally their disposal. In the extreme case of
products which in their normal use are introduced into
the environment in a concentrated or dispersed manner,
the emissions from production make only a very minor
contribution to the total pollution, which is caused
almost exclusively during use of the products. Integral
environmental protection must therefore also cover the
use and disposal of the manufactured products in order
to achieve a comprehensive "best practical environmental option". To avoid giving the impression of an

ideal world, it should of course be recalled that integral environmental protection is at present an objective and not reality. On the other hand, successful action is possible only if there are clear objectives.

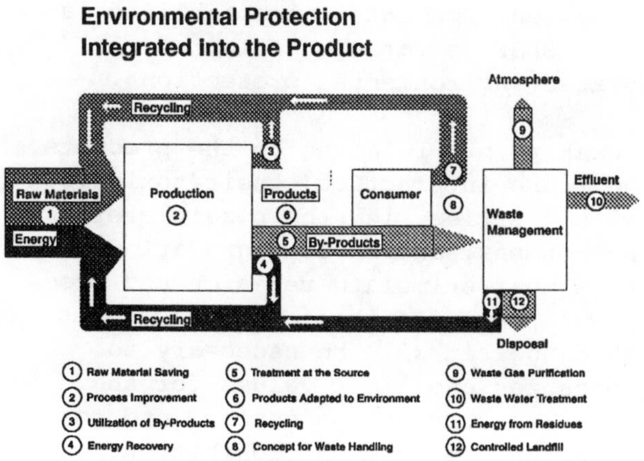
Environmental Protection
Integrated into the Product
Atmosphere
Recycling
Raw Materials
Energy
Production
Products
Consumer
By-Products
Waste Management
Effluent
Recycling
Disposal
1 Raw Material Saving
2 Process Improvement
3 Utilization of By-Products
4 Energy Recovery
5 Treatment at the Source
6 Products Adapted to Environment
7 Recycling
8 Concept for Waste Handling
9 Waste Gas Purification
10 Waste Water Treatment
11 Energy from Residues
12 Controlled Landfill

Figure 5 Environmental Protection Integrated into the Product

8 ECOLOGICAL OBJECTIVES IN RESEARCH AND DEVELOPMENT

In BASF, the environmental aspects of new investment projects at any site, in Germany or abroad, are investigated at an early stage of planning in consultation with experts on air pollution control, noise abatement, effluent treatment and waste disposal. From then until start-up, planning and implementation are appraised regularly, both at the drawing board and in the field. Before any decision is taken in relation to a capital expenditure - whether it be the choice of site for a production plant or the approval of funds for the initial or detailed planning of a new project, right through to the stage of final implementation - the BASF Board of Executive Directors requires expert assessments from the Environmental Protection and Safety Division. Funds are not approved until all outstanding safety or environmental problems have been clarified. This rigorous procedure ensures that conflicts of objectives are identified and clarified at an early

stage, before they can develop into major yes/no deci-
sions.

The search for better products never ceases. There are
always potential new products which are superior to the
existing ones; today, environmental acceptability is an
important criterion. BASF operates a policy of
"innovation integrated environmental protection".

The discovery of weak points relating to the products
with regard to their use in practice, their toxicological and ecological properties, ie. the possible effect
on man and the environment, and their degradation and
final fate in nature must stimulate research and product and process development. If these investigations
reveal a potential danger it will be necessary to
stipulate safety measures and limit values for the
relevant product and its further processing, and possibly also restrictions on use or even prohibition.

Calling for a substitute prematurely should however be
avoided. After all, the present products are the result
of a laborious development process; they have been
carefully tested with regard to benefits and risks. As
a rule, there is no substitute which retains all the
advantages and avoids all the disadvantages. Frequently, it is just that the substitute has been less
carefully investigated.

9   ENVIRONMENTALLY SAFE PRODUCTS

One of BASF's key research areas is the reduced use of
solvents in coatings and adhesives. These solvents
evaporate during use, ie. are released quantitatively
into the environment. Three developments provide a rem-
edy here. In water-based dispersions, organic solvents
are replaced by water, while in powder coatings and ra-
diation-hardened coating systems solvents are no longer
needed.

Polymeric detergent additives prevent lime deposits on
laundry and on washing machines. These incrustation in-
hibitors make it possible to dispense completely with
the use of phosphate in detergents and thus help to
protect surface waters from excessive eutrophication.

Furthermore, surfactants which pass into surface waters should be readily degradable. The indispensable "surfactant" constituents of detergents and cleaners must on the one hand be highly effective against dirt and grease in the washing process and, on the other hand, must produce little foam and undergo rapid and complete biodegradation in waste water treatment plants and in surface waters.

Fibreglass-reinforced vinyl ester resins are even superior to high alloy stainless steels with respect to corrosion resistance. They can therefore be used where materials are exposed to extremely corrosive conditions, for example, for the scrubbing towers of the stack gas desulfurization unit at the BASF power plant.

Catalysts assist· in carrying out chemical processes economically, ie. in increasing the yield, saving energy and hence for protecting the environment. BASF has been developing and producing catalysts for more than 70 years. Altogether 140 types are used for a very wide range of applications. New special catalysts for removal of nitrogen oxides from power plant exhaust gases and also traces of the famous dioxins from off-gases have now been included in the range.

10 INNOVATION, THE VITAL FORCE BEHIND INDUSTRIAL DEVELOPMENT

The chemical industry is a branch of industry with a particularly high innovation potential. As a rule of thumb, more than 50 % of its turnover is achieved with products which are less than 10 years old. This rejuvenating force also provides the impetus for the rapid improvement of environmental protection. It is not least for this reason that innovation must be continued, but danger threatens from three directions.

First, innovation must not be smothered by regulations, either in material terms, ie. relating to technical restrictions, or in terms of time, such as delay in publication, lodging of objections, explanations and administrative procedures through to approval. Valuable as the laws, regulations and specifications are, it is necessary to bear in mind that they are also obstacles

to innovation, and common sense must be shown. What is
required in practice is not to have as many controls as
possible but as many as are necessary.

Secondly, innovation requires time. It is an initially
exhaustive process, followed by technical procedures;
research, development, approval, ordering, erection and
commissioning are processes for which there is no short-
cut. As a rule, a new development takes years. Today,
environmental improvements are generally based on the
principle of precaution. There is no need to reject
reasonable deadlines for carrying out such improvements.

Thirdly, innovation also involves the acceptance of
risk. If rapid technical progress is desired, it is essential to have in mind that success in achieving new
developments can not be guaranteed. A pioneering process does not function at the press of a button but must
be optimized gradually. The necessary time must be allowed.

Environmental protection will continue to be a field of
activity in the future. The problems must be clearly
recognized and priorities must be defined unambigu-
ously.

Shell Chemicals UK's Response to the Environmental Challenge

J. F. Draper

MANAGER HSE, SHELL CHEMICALS UK LTD, HERONBRIDGE HOUSE,
CHESTER BUSINESS PARK, WREXHAM ROAD, CHESTER, UK

1 INTRODUCTION

The paper outlines Shell Chemicals UK's (SCUK) response to the "Environmental Challenge", first addressed following a vision exercise conducted by its Chemical Sector in 1990.

An environmental review was produced in parallel to SCUK's 1991 Country Business Plan which included detailed implementation plans and milestones by business and manufacturing location, costed emission reduction programmes, the first initiatives on Product Stewardship and manpower implications on research and development resources.

Emission reductions to date have been achieved by a combination of end of pipe solutions and the introduction of new processes. The paper includes brief descriptions of processes which the company is replacing or has replaced to effect considerable efficiency improvements and a consequent reduction in waste production.

Subsequent reviews have revisited the reduction programme and placed increased emphasis on Product Stewardship activities.

Continuous improvement aspects of our environmental programmes are emphasised in SCUK's contribution to the SUK Environmental Report (1992) which was included in the Shell UK Review. This covers all business and social reports for Shell UK and is published quarterly. The 1992 review represented SUK's first efforts to communicate environmental performance to the public using quantified emission data based on existing records.

2 THE SECTOR VISION

In 1990 the Chemicals Sector of Shell carried out a stra-
tegic review of its operations with respect to the impact
on them of environmental legislation and Shell's own
ambitions.

This led to the production of the "Environmental
Challenge" statement which was issued to all operating
companies. They in turn were requested to produce a quantified response which would be incorporated into their
planning cycles and was intended for update and review on
an annual basis.

The challenge asks a few key questions:

*   By 1995 develop a portfolio of products and associated
      operations in which the sector is confident of achiev-
      ing environmental excellence and long-term profit-
      ability.

It was, and remains, our belief that we will be in the
best position to achieve excellence if we are in
businesses in which we are profitable enough to provide the necessary levels of protection to achieve
this goal. Operations which are unprofitable rarely
operate in a manner which is consistent with caring for
the environment. One need only look at what has
happened in Eastern Europe to illustrate that fact.

In Shell Chemicals UK we believe that we can demonstrate that we have had regard for our impact on the
environment, even though we, in common with the rest
of the petrochemical industry, have had marginal and
cyclical returns on our investment in the past 10-20
years. A few instances of the way in which we have
reduced our impact are included in this presentation
paper, and we intend to focus on profitable operation
to improve our performance even further.

To clarify one point, however, in defining "By 1995 to
have developed.....", we do not intend to convey the
message that we will only be operating in profitable
business by 1995. Although we would clearly prefer
that to be the case, we recognise that making deci-
sions and carrying them out in an effective manner,
may take longer than that. What we do intend to convey is that by that date we will have completed our
studies, defined those areas in which we intend to
operate in the medium term future and may have begun
the process of exiting those businesses which do not
meet our criteria.

*   "Achieve excellence through the application of Product
      Stewardship...."

Product Stewardship is by now familiar to many. It is not an area which is dwelled on in this presentation; its principles and practices are being institutional- ised into our operating practices throughout the life cycle of our products and in all our dealings with our suppliers and producers.

* "Achieve excellence through progressive improvements in our operations with the ultimate aim of eliminating by virtue of such continuous improvements, emissions, effluents and discharge of materials that are known to have a negative impact on the environment."

SCUK's response is the substantive part of the paper since it deals with the approach that we are taking and have taken to address this part of the challenge and is essentially directed at waste minimisation.

3 HISTORICAL PERFORMANCE

In SUK and SCUK, the monitoring and reporting of our emis- sions has been a regular feature of monthly, quarterly and annual internal reports for some time. Our management by objectives system with its definition of responsibilities has ensured that line management recognise those parameters which are judged to be important and act to minimise or reduce them. Last year for the first time we issued infor- mation to the public on our emissions in the UK in the Shell UK Annual Review. This details emissions from all our major Expro, Downstream Oil, Chemical and Metals opera- tions and reflects our improved performance.

4 SOLID WASTE

We have in the UK followed a policy of minimising waste to landfill for some considerable time. In 1978 we built an energy recovery unit at our Stanlow refinery to incinerate waste generated at our facilities and generate steam from the heat released which we return to Stanlow refinery for use in its processes.

This facility treats material which could otherwise be disposed of directly to landfill, but would be of higher volume and contain, in our view, a higher long-term envir- onmental risk.

For Chemical operations, this reduction has been achieved mainly by optimising reaction conditions and concentrations of bleed streams on processes which have been in operation many years, and demonstrates that major improvement can be achieved on even the oldest processes.

A spin off of the reduction at Stanlow is that it has reduced the total amount of waste being sent to the energy recovery plant. This reduction has meant that the expen- diture planned to debottleneck this facility, millions of pounds, has been cancelled.

5 LIQUID WASTE

Much of our liquid waste is incinerated on our own facilities at Stanlow. We dispose of all aqueous effluent by
segregating from water insolubles by gravity in an inter-
ceptor (API or CPI) and then pumping the remainder to a
biotreater where the residual load of the effluent in the
aqueous environment is substantially reduced prior to discharge to local water courses.

Shell funded the building of the local biotreater for
its chemical effluent at Stanlow and now has its effluent
treated there and pays a fee which is a function of volu-
metric flow and contaminant in the effluent.

Between 1988 and 1992 we reduced effluent load by
almost 50%. This reduction is a result of a combination
of shutting down operating plants in 1992, and by the steady
reduction in effluent load in previous years due to
improvement in the operation of facilities built there
between 1960 and 1980.

The improvements have been made by modifying distillation column control conditions or reaction control parameters to improve yield. None required the expenditure of
significant capital for individual projects, as was also
the case with our solid waste improvement projects, but for
example, the total amount spent on HSE improvements at
Stanlow, initiated by our process control, technological
and chemical investigation groups, amounts to £1-2M/annum,
demonstrating that continuous improvement has been an
essential part of our philosophy for some time.

Our efforts at reducing effluent loading from our
operations to the biotreater have made a significant impact
on its operation, so that further capex to install additiona
capacity has been cancelled and a review is being made of
the potential for biotreating other waste streams, not
currently biotreated, on this facility.

6 GASEOUS EMISSION

Our gaseous emission story is somewhat different. We
routinely target and track our emissions performance against
sulphur dioxide, carbon dioxide, nitrogen oxides, volatile
organic compounds and halons.

From our chemical operations, our emissions of these
substances are low, and certainly low compared to the rest
of Shell's operations in the UK, and so I would prefer to
concentrate on a parameter of gaseous emission which we
also track to improve our performance.

The gaseous emission considered is any one which produces a complaint from an external or internal source. Each of our sites operates a complaints procedure and records and investigates, with the public, if that is the source of the complaint, any complaint it receives.

In 1987 we registered 13 complaints, rising to 27 in 1989, and reducing to 8 in 1991.

We believe that the initial rise was a typical reaction similar to that demonstrated whichever quality systems are introduced. We believe that the overall reduction, 13 to 8 in this period, is a very noteworthy achievement since the public's expectation has been raised to a level where reports are received which we believe they would have ignored and not reported in the past.

We also track smoking across the refinery with cameras which record the emission from chimneys and flares and we sum the smoking time and record it. In 1987, the sum of refinery chimney or a flare smoking time was about 12 hours and in 1991 this was reduced to about an hour.

Some of these reductions were the result of changes in fuel, but the majority were again effected by implementing the recommendations of our supporting technological groups changing control and start-up routines with relatively small amounts of expenditure.

So, this is the background of how we've been managing to date. The 1992 SUK review contains details of these and other emissions that we have tracked.

In 1991 we assessed emissions from our chemical operations and developed plans for emission reductions in the period 1992 - 1997.

In 1991/2, the Environmental Protection Act and detailed knowledge of the requirements being made on individual process operations as a result of HMIP enforcing Integrated Pollution Control, were not clear.

This legislation requires that an emission list is lodged on the public register, and that emission levels are agreed with HMIP which satisfy BATNEEC or will be modified to achieve BATNEEC requirements by the end of the licence period for that process.

In 1991, trying to predict the effect of this legislation on our operations was difficult, so we made our best estimate of the requirement and produced emission reduction estimates.

In parallel to this exercise, we identified areas where we could, either at low cost or cost effectively, implement capex programmes which would have significant impact on reducing our overall emissions.

Overall reduction levels were calculated at 30 to 40% at each location, but the potential for achieving significant reduction varied from plant to plant.

We will achieve this by continual review of processes, improving their control and optimising the way they operate: an extension of the sorts of small changes we have been implementing to achieve the reductions we have made so far. But, clearly, apart from areas where grade changes are being made, our ability to optimise processes must be reducing. This is certainly true in solid and aqueous waste reduction where we have been able to accurately measure our emissions for some time. However, in gaseous emissions there are still some areas where we can effectively reduce emissions as we start to become better able to quantify them. We can look at ways to reduce our gaseous emissions and measure the effect of the changes we have made.

The calculations and methods used to assess VOC emission from process operations has been historically a difficult and rather inaccurate operation. Presently, some believe that methods used also for estimating, say, breathing losses from tanks or other fixed pieces of equipment have been inaccurate. Anyone who has attempted to close a material balance around an operating unit, especially one containing liquified gases, knows the difficulty of closing such a balance. This uncertainty has made it difficult to estimate losses, and consequently to justify changes and prove their worth except in some special circumstances.

We are continuing to do things in the coming years which are consistent with reducing emissions from known loss areas, like tanks and process venting, but we are investigating ways of better quantifying emissions using OPSIS and LIDDAR techniques. We would like to be able to point an instrument at a refinery gaseous emission bubble and measure the sum of emissions from that operation. With more confident knowledge of the relative magnitude of the sources, we should be able to confidently move to gain maximum reward for our efforts. We are currently testing our own equipment and are co-operating with the National Physics Laboratory to field trial their instrumentation, but our view is that there is still a great deal of work to be done to validate the results from such instruments. However, we believe that, like the cameras we have used to track the 'smoking' performance of our operations, these instruments will be essential for continuous monitoring of our gaseous emission bubbles.

We are also considering the use of some 'bolt-on' end pipe solutions for specific problems. One such potential solution is in NOx reduction for effluent gas systems.

Shell has, through its own research efforts, identified a system for reducing NOx emissions in fluegas exhausts - the Shell DENOX system. This system comprises a Shell designed catalyst and reactor system which can be retrofitted in a relatively simple and low cost manner to flue gas streams from, for example, ethylene crackers or gas fired heaters and boilers to reduce NOx emissions.

The system consists of a catalyst contained in a wire mesh type reactor housing where, in the present of ammonia, the NOx is converted into nitrogen and water. The catalyst is a mix of Ti/V on a substrate and the reactor configuration can be tailored to the application's specific operating requirements such as NOx conversion, ammonia slip, pressure drop and plot space. A major advantage of the system we are looking at is that due to the intrinsically high reactivity of the catalyst, high NOx conversion levels can be obtained at temperatures below $220^{\circ}C$. Typical operating range is $120^{\circ}C$ to $350^{\circ}C$.

In the conventional line up for NOx removal the catalytic system operates immediately after the heater at temperatures around $350-400^{\circ}C$.

Typical operating data from some installed equipment is:

Firstly, at the site where our catalyst is made we have installed a NOx removal system in the off-gas from the process. NOx removal of 98% has been achieved, reducing flue gas concentrations from 11,000 mg/Nm^3 to <200 mg/Nm^3.

Secondly, performance data for an installation on flue gas from process heaters operating at $180^{\circ}C$ **shows** NOx levels in the final effluent gas **have** been reduced to 10 mg/Nm^3, with a removal of more than 88% of the NOx present.

From examples of an end of pipe possibility for effecting a reduction in our emissions, I should now like to move onto an area where we are replacing an existing process, and again, achieving substantial reductions in our total emissions.

7 ADDITIVE MANUFACTURE

One of the plants at Stanlow produces overbased salicylate products which are used as additives in lubricating oils. All overbased products are produced from salicylic acid in a multistep process which has been operational for some 20+ years.

The basic process steps of the process are:- Olefins and Phenol are reacted together to form alkylphenols or alkylate. The alkylate, when refined, is first reacted with caustic soda to form sodium alkyl phenate, then reacted with carbon dioxide to form sodium alkyl salicylate or carboxylate. Finally, the dope acid, salicylic acid, is "sprung" by the addition of sulphuric acid.

The process had been in operation for a long time and operated with local venting, and disposal of waste streams in a manner which we knew could be improved to bring it up to the latest HSE standards and considerably reduce emissions to the environment.

The primary area for improvement was the alkylation section of the process.

A material balance of the process showed that some 420 tpa of cake was disposed of to incineration, with the residue post incineration sent to landfill. This was expensive and inherently hazardous with the potential exposure to phenol both in the process and the transport of the residue.

A new alkylation process has been developed at KSLA - Shell Research labs in Amsterdam - in which the olefin/phenol mix is passed through fixed beds of catalyst to effect the desired conversion. The resulting crude alkylate is solid free and requires no filtration and can be immediately distilled to recover the lost phenol.

As the catalyst is used it slowly de-activates and is regenerated by burning off the laid down carbon products on the catalyst with a hot nitrogen stream containing controlled amounts of oxygen to produce oxides of carbon and water only. When the catalyst loses its activity it can be returned "clean and dry" to the catalyst manufacturing or precious metal recovery contractor for treatment and recovery before being disposed of to landfill.

The reduction in waste streams in the section to 80 tpa from 420 tpa and the prevention of any loss of phenol with the consequent reduction in associated risk represents a major improvement. We have made similar improvements in the other stages of the process. Our estimate of the total potential savings is 1935 tpa of waste streams reduced to 310 tpa, a considerable achievement

8 POLYMER PRODUCTION

A further area where we have achieved reduction is on the polypropylene production plant at Carrington. Our original plant at Carrington was a 100,000 tpa homopolymer plant which used the slurry process to effect conversion of the liquid propylene to polypropylene. It was brought on line at Carrington in 1973, and in it propylene is reacted with a slurry of polymer and solvent. The residual process blocks first remove and recycle excess propylene and then the polypropylene is treated to decompose the catalyst, then recover and clean up the solvent to make it suitable for recycling.

In the late 70's Shell Research Labs in Amsterdam developed a process, the LIPP Process (Liquid Propylene Polymerisation Process) in which the polymerisation was carried out in liquid propylene, and it has recently made another major advance by developing the so-called SHAC catalyst (Super High Activity Catalyst) which has very high stereo specificity.

The combined process, LIPPSHAC, offers considerable advances on the slurry process and a 135,000 tpa plant was built at Carrington and brought on line in 1992.

The new process offers major process advantages since:

* Catalyst decomposition, slurry washing, solvent removal and recovery and the drying steps are no longer required.

* The losses of polypropylene to unwanted atactic polymer is reduced from its traditional production rate of 2-3% of intake.

* Utilities consumption is significantly reduced.

* Process chemicals are no longer used and decomposition products are avoided.

A principle benefit is in the reduction in energy usage. In the slurry process, we used 10.2 Gj of electricity to produce 1 tonne of polypropylene. The new process uses 2.5 Gigajoules, a saving of 75%, with a beneficial reduction on emissions to the environment and the consumption of natural resources.

And what about the cost of all these measures? Without providing a comprehensive review, there is one interesting effect which was reinforced by our cost analysis. If we calculate the ratio of capital costs of achieving reductions by the difference of the emission tonnages achieved it is clear that we may be able to make significant reductions in some of our processes for a relatively low cost per tonne, but this low cost per tonne effect is generally valid when there is a significant reduction being achieved. Consequently it still requires significant total expenditure. Smaller tonnage reductions on other existing units are achievable at high cost per tonne of emissions reduction. If we prioritise, where should we spend our money first?

We need to consider the implications of the analysis on our target of continuous improvement. It infers that as we try to reduce costs on processes with low levels of emissions, the total cost per tonne of reduction will go up. So, if it is expensive now, it is going to be even more expensive in future, and we need to develop tools which allow us to better evaluate the cost/benefit relationships for our emission reduction programmes. Reduction to zero will indeed be costly, and will it be beneficial?

9 POLYMERS WASTE MANAGEMENT

One of our activities which is outside our refinery, but which demonstrates how our product stewardship and waste minimisation programmes interact throughout the life cycle of our product is polymers waste management.

We are keen to ensure that the polymers we produce are handled in the consumer chain and thereafter in a manner which is consistent with minimising the generation of waste.

There is no longer any argument that whilst disposing of polymers to landfill is perfectly safe, it is wasteful of their potential value as secondary raw materials or as energy. Thereafter, the debate is complex regarding the best way to realise that potential.

Shell, together with many in the environmental movement, now promote the 4 R's or the hierarchy of waste management when considering the final fate of their polymeric materials.

REDUCE   by avoidance if possible

REUSE

RECYCLE

RECOVER   as raw material or energy

At the heart of the debate is the widely held view that recycling is good and energy recovery means incineration and is therefore bad.

The Shell group has devoted time and effort to the debate of this issue, and, without dwelling on those arguments, it may be fair to say recycling means many things to many people, and may be good or bad, depending on specific circumstances. Is recycling good if it consumes more energy than the production of virgin materials? Similarly, energy production either from traditional fuels such as coal, oil or gas or from waste, can be good or bad depending on the technology and the management of its specific facility. Finding the best balance between these in terms of environmental benefit and overall cost to society is a key issue and will surely remain one for some time to come.

If we do any of these, instead of disposing of our material to landfill, we minimise waste and the consumption of natural resources.

Shell, as well as remaining active in the debate, is looking at the potential for recovering the waste as raw material in a project at its research laboratories in Amsterdam.

Converting plastics back into the oil feedstocks from which they originated is a potentially appealing concept as many of the quality-technical limitations on finding a market in competition with virgin polymer are removed. It is also likely to show a better energy balance than incineration as hydrocarbon molecules are not destroyed. Public acceptability is also likely to be much better as plants can be located and seen as part of existing refinery complexes. However, there are still some major hurdles to be overcome. Firstly, logistics - petrochemical plants must be very large in order to achieve economy of scale and so collection and sorting of sufficient plastics will need to focus on areas of high population density, ideally near to oil or petrochemical sites. Secondly, fully developed technology is not yet available although research has been in progress for the past 2-3 years and several routes are now looking promising.

A key concept in most of these technologies, including the project being progressed by Shell Research Laboratories Amsterdam, is that of a "stand-beside" feed preparation unit which would convert mixed plastics waste into an oily feedstream suitable for co-feeding to existing petro-chemical or refinery plants for additional distillation, gasification (e.g. Shell SGP process) or hydrogenation into more refined products. During this conversion process, impurities and non-hydrocarbons such as chlorine are removed to ensure compliance with normal feed specifications. Based on logistic considerations plants of about 50 Kt/a could be envisaged which in turn would feed into typical 500 Kt/a user plants. Whilst some ad hoc feedstock recycling initiatives have already been taken in Germany, it is unlikely that any major new commercial plants will be operational until about 1997. This is based on the need for a further 2-3 years R&D work, including pilot plant construction and operation, plus a further 1-2 years commercial plant construction time, i.e. 3-5 years in total from today.

So, how much will feedstock recycling cost and what proportion of the waste problem might be covered? Probably the major cost for feedstock recycling will be the incremental collection, cleaning and shredding costs compared to landfilling or municipal incineration. The stand-beside concept for a feedstock feed preparation unit minimises the dedicated capital expenditure and utilises, as much as possible, the enormous economy of scale of existing refinery or petrochemical user plants.

Savings might be possible in collection if an integrated waste collection and separation plant could be employed rather than separate collection of plastics. However, these savings might be offset by the need for greater attention to subsequent clean-up. Overall costs for the stand-beside "converter" process are seen to be about half those for a stand along plant but are significantly higher than for municipal waste to energy plants.

So, that's some of the background to some of the activities that we have taken to respond to the environmental challenge at Shell and reduce the generation of waste in our facilities.

I hope that I have demonstrated the seriousness and commitment of our approach, but in finishing, let me draw your attention to one of the other statements in our policy.

Shell Chemicals UK recognises the concerns of employees, shareholders and society in general on environmental protection matters and provides them with information to explain the company's policies, practices and performance against targets.

A Mori Poll in 1991 showed that the public's opinion of the UK chemical industry was at an all time low. One of the few industries perceived less favourably by the public is the nuclear industry.

At the same time as our public image has deteriorated, the number of environmental laws adopted by the EC each year since 1976 has increased from 1 or 2 to 20 or 30 per year. We may draw our own conclusions about any connection between the two factors, but I believe that they tell us that if we don't improve our performance and the public perception of us, we risk losing our permit to operate.

This paper illustrates how in the Shell UK review and in other ways we have started to provide data, but another vehicle in use to help us improve on both counts is Responsible Care, which was introduced into the UK in 1989 by the Chemical Industries Association (CIA). It is an initiative adopted by the chemical industry and it is compulsory for all members of the CIA to adhere to the principles of Responsible Care.

It is world-wide and intended to improve the performance of the chemical industry in the field of health, safety, environment, product safety, distribution and relations with the public. It is designed to enable companies to demonstrate to the public that these improvements are not a public relations exercise, but if we improve and don't tell the public, can we hope to change their perceptions? There are many parts of Responsible Care and everyone should be encouraged to become familiar with it and to take on dialogue with the public as a routine part of their job in the industry.

The Mori Poll for 1992 shows no further deterioration and offers us some hope that we can change things. It is up to all of us in the industry to help maintain this trend by continuous improvement and dialogue with the public.

A Corporate Commitment to Waste Minimisation

R. E. Chandler

MONSANTO PLC, CORPORATION ROAD, NEWPORT, GWENT NP9 0XF, UK

HISTORY OF MONSANTO COMPANY

Started in St.Louis, Missouri, in 1901 when a pharmaceutical company employee called John F. Queeny decided to start up his own business naming the company after his wife, Olga Monsanto - and became the first US manufacturer of Saccharin.

Monsanto has 30 manufacturing sites and 30,000 employees. Sales in 1992 were nearly $8 billion. It is the 14th largest chemical company in the world. The corporate headquarters is at St. Louis, Missouri, with 4,000 staff.

MONSANTO IN EUROPE

Monsanto and its subsidiaries employ about 4,000 people with 6 plants, 2 technical centres and sell products for almost $2 billion annually. The geographical spread is quite wide.

There are sales offices in almost all European countries, including Eastern Europe and the former Soviet Union.

The European headquarters is at Brussels employing 400 people.

These are three major manufacturing sites in Britain: two in Wales and one in England.

Monsanto's £ 40 million plant at Newport in South Wales was started in 1949.   It employs 300 staff to produce about 60,000 tonnes of product each year.   Main products are phosphonates for the detergent and water treatment industries; plasticisers; heat transfer fluids; rubber chemicals; and thermoplastic elastomers (joint venture with Exxon).

Monsanto, Ruabon, near Wrexham in North Wales produces about 40,000 tonnes of product each year.   400 people are employed in making rubber chemicals and detergent chemicals.   Because the Ruabon plant is a semi-rural location in close proximity to a number of villages and is next to the river Dee which is used for fishing and water extraction, special emphasis is put on· environmental protection.

G. D. Searle, Morpeth, Northumberland, employ over 500 people at the site which was officially opened in 1970.   Products are ethical pharmaceutical and over-the-counter products in tablet, capsule, liquid and powder form.   Cytotec is manufactured for prevention of certain ulcers.

CORPORATE POLICY

Environmental policies and action programmes in Monsanto have been evolving over the last 20 years.

Although we feel we have made significant progress towards waste elimination we recognise that we are aiming at an ever more difficult target and that we still have a long way to go.

Over the last 2 decades or so, environmentalism as a force at both national and international political levels has grown strongly. Over the same period of time the chemical industry has responded to environmentalism in what can be perceived to be three distinct but overlapping phases.

The first phase can be described as the 'denial' phase when confrontation was the order of the day and when industry tended to deny that major environmental problems existed and argue that new regulatory programmes and 'massive' government intervention was not necessary.

Characteristic of the "denial" phase was that industry would predict dramatic high cost and unacceptable economic impact in terms of widespread factory closures, job losses and export of the jobs and industry to less developed countries.   Such exaggerated forecasts of economic doom and gloom cost us a lot in terms of credibility with legislators and the public in general.

The second phase may be called the 'risk/benefit' phase.   This phase could be characterised as confronting emotions with reason, and fiction with scientific facts.   Major chemical companies, particularly in the USA, launched expensive programmes in an attempt to 'educate' the public opinion formers by using all the communication media available to them.   Monsanto's "Chemical Facts of Life" was such a programme in the late 70's and early 80's.   That certain low risks may be 'acceptable' when associated with significant benefits was, and still is, a very difficult concept to successfully argue.   Who is to be the judge of what is 'acceptable'?   Public emotions and fears of the unknown did, and still can, win over what we in the industry view as facts and reason.

The third phase is termed the 'public-right-to-know' phase.   It is debatable that this phase was triggered by the infamous Bhopal disaster in which 3,000 people in the vicinity of a chemical factory in India were killed.   This tragedy shook the chemical industry to its foundation. It was the "worst case" scenario come true.   Looking back at the aftermath of that tragedy, it is apparent that it was the catalyst for a new and unusual response from industry.

This approach meant initially a willingness to disclose environmental data to plant communities and other members of the public when asked to.

Today it means more than passive openness and includes active efforts to establish connection with local community groups, with whom we not only pass information to but listen to and try to understand their concerns about the possible impact of our operations on their everyday lives. Such proactive openness is a two-edged sword. On the one hand it hopefully creates public trust in operations, but at the same time, it will generate it's own pressure for improved performance. In other words, we are inevitably in the situation of having to earn the right to operate from our communities. This may in many cases mean that it is no longer sufficient to obtain an official operating license from the relevant authorities. The so-called 'public-right-to-know' concept is very alien compared to what we have been used to. The CIA 'responsible care' programme is a step change from the preceding attitude.

As public concerns about the environment have grown over time, so has Monsanto's efforts to minimise its adverse impact on the environment. During the early years this effort and the resources allocated to it tended to be somewhat ad hoc and local in character.

<u>CORPORATE POLICY</u>

In the mid 1970's Monsanto developed a philosophy relating to corporate responsibility. It related to issues such as equal opportunity, health and safety, conservation of energy and natural resources, and environmental protection. These global aims were translated into more specific policies and targets. Then in 1977 a significant change occurred in the company's approach to environmental management. This was marked by the appointment of a senior member of the board of directors who became responsible for corporate environmental policy and its implementation worldwide. From that year on, individual operating units of Monsanto have had to demonstrate their individual achievement performance and against comprehensive targets spelled out in a set of detailed worldwide management guidelines:

1. **EFFLUENT AND EMISSION CONTROL**
   Reduce pollutants in effluents and emissions from Monsanto operations to meet corporate targets, going beyond those levels either required for regulatory compliance or necessary to protect health and the environment.

2. **WASTE MANAGEMENT**
   Design and operate facilities to minimize in waste streams the routine and accidental release of pollutants to the environment. Over the long term, work toward the ultimate goal of zero released to all media. For wastes that remain, use waste disposal practices that achieve compliance with regulations and which achieve acceptable environmental impact, no health effects, minimum long-term liability and cost effectiveness. Continue waste management programs that establish Monsanto control of disposal and that favour alternatives to land disposal.

3. **PLANT ENVIRONMENTAL ASSESSMENTS**
   A program of environmental assessments and audits of all plant sites will be maintained in order to assure regulatory compliance and the protection of the surrounding environment.

4. **EMPLOYEE AND COMMUNITY SAFETY AND HEALTH**
   Monsanto will provide a healthful and safe environment for its employees and community neighbours and will monitor and evaluate employee health status, determine and monitor

workplace factors affecting employee safety and health,
comply with Monsanto workplace exposure guidelines and with
government safety and health regulations, review and major
capital projects to protect the health of people at work and
in the community.

5.   **OUTSIDE PROCESSORS**
Select companies for support of Monsanto operations –
through product conversions, custom manufacture,
formulating, by-product sales, waste management, and other
services supporting Monsanto businesses – which will operate
with concern for worker safety, regulatory compliance,
community protection and protection of the environment.

6.   **PRODUCT STEWARDSHIP**
Monsanto products and intermediates will not present an
unreasonable risk of harm to human life or health, or to the
environment when they are properly handled, transported,
used or disposed. Stakeholders will be provided information
regarding handling, storage, use and disposal of Monsanto
products.

The fundamental principle of these guidelines was that legal
compliance is only the minimum requirement and that we must impose
stricter demands on our operations if dictated by additional needs
for environmental risk reduction.

These required an assessment of the impact of all of our
discharges.

To assess the impact of releases we needed to quantify them. In
Britain certain licensed processes have been controlled for many
years, but the gaseous emissions have generally been related to
ground level concentrations and personal exposure rather than
environmental protection. In West Germany quantitative limits on
emissions have been introduced since the late 1960's and as
Monsanto opened plants progressively on the continent from that
time all our stack emissions and aqueous discharges have been
subject to tight controls. TA LUFT is incorporated into Belgian
law.

AIR EMISSIONS REDUCTION PROGRAMME '90% PROGRAMME'

In 1988, Monsanto's Chairman, Richard Mahoney publicly announced a corporate goal worldwide to reduce our air emissions reported under the United States Superfund Amendment and Reauthorisation Act 'SARA' legislation by 90% by the end of 1992, with respect to the base year of 1987. It was one of the most ambitious voluntary programs ever undertaken by an international company. The air emission goal was amended for Europe to include chemicals not just in the SARA title III list but which are subject to local concern or control. Examples are ethyl acetate, ethyl alcohol, carbon monoxide and sulphur dioxide.

By the end of 1992 we reduced air emissions of these chemicals by 92%.

To meet our target we first needed to know what our current emissions were. The data collection policy was extended in Monsanto to cover all sites worldwide. Gathering and reporting the data caused a lot of turmoil within the Company because it required much effort. It has also had several benefits; it has exposed gaps in our knowledge of how our plants operate. Some of our assumptions on emissions were shown to be invalid.

Although we have always has accurate measurements of our solid and liquid discharges because of much earlier legislation, we realised that much of our air emission data was based on estimates. Because of impending new TA LUFT standards (established in Germany), our Belgian plants took the lead in embarking on a systematic programme of obtaining a complete analysis of every point source emission. They began to estimate fugitive emissions in a more scientific way by means of a dedicated mobile laboratory. Now these measurement programmes have been extended to our 2 sites in Wales.

AIR EMISSION MEASUREMENT

The first step is qualitative screening of potential emissions by use of a portable photoionisation detector. Any flanges found leaking are then attended to.

(OUTSIDE MOBILE LAB) The mobile laboratory is a purpose-built unit set up for maximum flexibility in enabling a wide range of air emission sampling challenges to be met.

(INSIDE LAB 1)   Current major equipment includes: Gas Chromatograph (G.C.) with 2 flame ionisation detectors, mass spectrometer, gas sample loop, split/splitless injection ports and catalyst, entirely under control of the lab. P.C.

Usually bag samples are brought to the G.C. but a 100m heated sampling line facilitates continuous real time emission profiles to be established. A multi-path length infrared machine is also part of the sample control system associated with the G.C.

(INSIDE LAB 2) The lab. has all the essential services, including heating, and is air conditioned. This is necessary as the technologist spends a long time in it. It has a built-in fume cupboard.   Amongst its portable equipment are various flow measurement kits,   custom sampling devices, ion-selective electrodes for inorganics and the photoionisation detector.   The P.C., through a wide variety of software, is capable of all data analysis and reporting.

A major benefit of this work has been that for the first time for many of our processes we have been able to complete mass balance calculations. This then allows yield and cost improvement work to be scoped.

Another significant benefit of the efforts to identify and quantify our fugitive and process emissions is that they have been of timely use for describing releases when applying for Integrated Pollution Control (IPC) authorisations.

PERCENT REDUCTION BY TECHNOLOGY

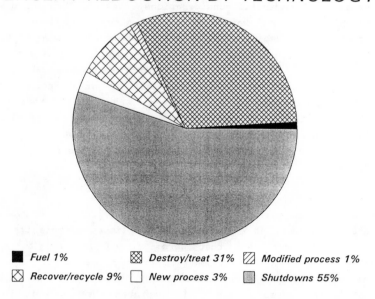

■ *Fuel 1%*	▨ *Destroy/treat 31%*	▨ *Modified process 1%*
⬡ *Recover/recycle 9%*	☐ *New process 3%*	▨ *Shutdowns 55%*

The above diagram shows just over one half of the emission reduction was attributable to plant closures. But no plants were shut for this reason. 'End of pipe' treatment e.g. scrubbing and/or condensing are common techniques for removing air emissions. At Newport for example we wanted to reduce formaldehyde (methanal) emissions resulting from tanker unloading. To do this we installed a scrubber which uses a weak sodium hydroxide solution to dissolve the HCHO.

At our Antwerp and Newport plants we have made significant use of incineration. We have installed 4 incinerators at Antwerp and one at Newport for large volume, low concentration streams where recovery is prohibitively expensive.

The goal set by Monsanto was one of the most ambitious set by any company of the chemical industry, both in terms of reduction and timescale. The tight schedule has meant that in some cases proven 'end-of-pipe' techniques have been used where development of source reduction would have taken too long.

I regret I can not give you details of most process improvements made because of commercial considerations. However I can give you two examples: In our Maleic Anhydride process at Newport we catalytically oxidized n-butane to maleic anhydride. Some of this was then scrubbed in water. The resulting solution of maleic acid (which contained the trans isomer; fumaric acid) was dehydrated using solvent-grade xylene to boil off an azeotrope. When studying xylene losses to atmosphere it was realised that each isomer forms a different azeotrope and so they vary in efficiency of dehydration and therefore product recovery. So we switched to a more expensive supply of xylene to reduce air emissions, and to save money.

Using a novel application of a technology called a cyclone separator, engineers at our Ruabon plant eliminated more than 55 tons of trichloroethylene that were emitted annually to air. Trichloroethene is a solvent used in product purification. The new device, designed by the Atomic Energy Authority, is about the size of a home deep freeze. It captures the chemical in a powerful vortex and then recycles it into the manufacturing process.

PERCENT CAPITAL BY TECHNOLOGY

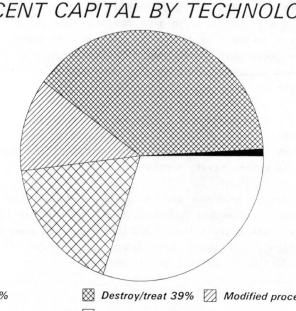

■ *Fuel 1%*   ▨ *Destroy/treat 39%*   ▨ *Modified process 12%*
▨ *Recover/recycle 18%*   □ *New process 30%*

Closures and shutdowns are done for business reasons and so the costs are not attributed to the air emission reduction programme.

Monsanto worldwide has implemented >250 projects at a cost of more than $100,000,000. A significant contribution towards our target has been made at our European chemical plants where 37 projects have been implemented.

The Monsanto Pledge

It is our pledge to:

- *reduce all toxic and hazardous releases and emissions, working toward an ultimate goal of zero effect;*

- *ensure no Monsanto operation poses any undue risk to our employees and our communities;*

- *work to achieve sustainable agriculture through new technology and practices;*

- *ensure groundwater safety;*

- *keep our plants open to our communities and involve the community in plant operations;*

- *manage all corporate real estate, including plant sites, to benefit nature; and*

- *search worldwide for technology to reduce and eliminate waste from our operations, with the top priority being not making it in the first place.*

Monsanto
January 1990

Richard J. Mahoney
Chairman and Chief Executive Officer
Monsanto Company

AVOID WASTE AND REDUCE EMISSIONS

It is the 1st and 7th pledge items which have led us to the A.W.A.R.E. programme: Avoid Waste And Reduce Emissions.

This programme is separate from the 90% air reduction one and goes further in that it includes releases to all media.   Any air emission projects which became operational from 1990 - 1992 will contribute to the new program.

The 1990 (base year) release total was 170,000 tonnes of which Monsanto, Europe released 8%.

The chemicals are the SARA (air), EC129 'black' or list 1 substances (water) and CO (air), but to all media.

For this program we are currently evaluating source reduction opportunities to reduce discharges of effluent.   I anticipate this will put us in a knowledgeable position when discussing BATNEEC with HMIP (H.M. Inspectorate of Pollution) for authorisable processes.

These pledge guidelines replaced the six environmental, safety and health (ESH) guidelines described earlier.

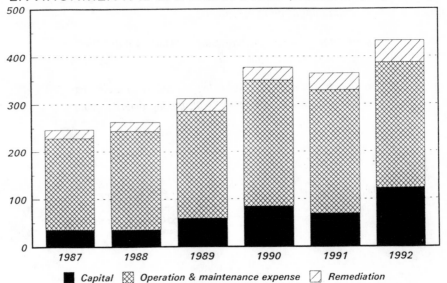

ENVIRONMENTAL EXPENDITURES (millions of dollars)

The above graph shows a progressive increase in Environmental spending for site clean ups, capital investment for emission reduction and operating costs.

In 1992 expenditures were approximately $123 million for environmental capital projects and approximately $264 million for operation and maintenance of environmental protection facilities.

The Financial Times of June 18th, 1993 listed 20 chemical companies and their environmental expenditure. Monsanto ranked third in this FT survey with expenditure of 5.5% of sales.

Monsanto estimates that during 1993 and 1994 approximately $75 million-$125 million per year will be spent on additional capital projects for environmental protection.

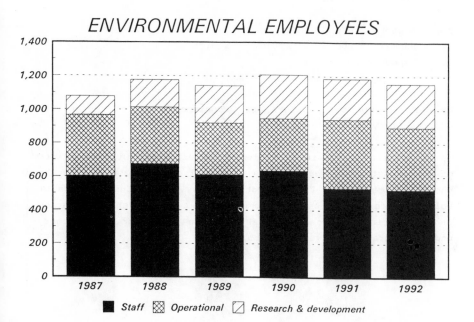

In 1992, there were about 1,155 employees involved with our Environmental, Safety and Health programmes.   At each manufacturing plant there are people who are responsible for ensuring compliance with regulatory and corporate requirements, who analyse waste streams and operate waste treatment and disposal facilities.   There are research centres in the U.S. and Europe and corporate staff who provide a core of expertise and audit site performance.

FUTURE

The current programs in Monsanto and industry in general have been mass based.   These equate one tonne of CO with (say) one tonne of benzene.   Clearly there is a difference in effect on the environment and people and so we want to rank chemicals.   We envisage our next programme will use an effects based model including the aspects of:

1.   Acute and Chronic effects.
2.   Mammalian and Environmental toxicity.
3.   Exposure potential.
4.   Ranking of chemicals.

We will attempt to structure our programme to achieve scientific acceptability outside Monsanto.

We anticipate increasingly stringent requirements will be placed on our industry.   Todays voluntary proactive stance may one day be the statutory norm.   This is not far away.   H.M. Inspectorate of Pollution describe the requirements of a 'release minimisation programme' in the recent process guidance notes for our industry under Integrated Pollution Control (IPC).

Waste Minimisation – A Small Company Perspective

B. W. Trenbirth

CONTRACT CHEMICALS (KNOWSLEY) LTD, PENRHYN ROAD, KNOWSLEY
INDUSTRIAL PARK SOUTH, PRESCOT, MERSEYSIDE L34 9HY, UK

1. LEGISLATION

The implications of the legislation which has resulted
from the Environmental Protection Act have widely
affected the Chemical Industry. In particular we are
concerned with waste minimisation, an important
cornerstone of the legislation. As can be seen from
Environmental Policies throughout the industry, nearly
all companies have a high level commitment to
environmental control, but how is this commitment put
into practice? The concern is that we may just be
paying lip service to the law without benefiting from
the many payoffs that an aggressive Environmental
Policy will give.

The implications of the introduction of Integrated
Pollution Control are that companies will tend to
specialise more as their hard won authorisations come
into force. By the time we see all existing processes
under the control of HMIP we could find companies with
fewer areas of expertise as small processes fall by the
wayside. The advantages are that we should see more
resources being applied to fewer areas with a subsequent
commercial advantage resulting from more efficient and
less wasteful processes. This gives us commercial
advantages not only in the U.K. but also against
foreign competition as the environmental legislation
spreads slowly across the world.

The introduction of BATNEEC should be a
significant opportunity for us to exploit the
commercial implications of IPC. BATNEEC is to be
applied using the following criteria:-

1. New and cleaner processes

2. Reduction and recycling techniques

3. Treatment

These are the guiding principles in order of preference published in the Guidance Notes to IPC. What they require is that the primary aim is to assess the process routes available for both new and existing processes. Implicit in this review is a justification of the solvents used and an in-depth consideration of the potential recycling methods that may be available. It would be wrong to only look at chemical technology; we should also look at the many physical techniques available that can help with waste minimisation. Only when all other aspects have been considered should we consider the treatment of the effluent produced. The cost involved in this exercise is considerable but we should see a considerable commercial advantage as well as an environmental benefit.

It is implicit in the legislation that all authorisations will be reviewed every four years. It is clear that the best techniques available at that time will be required to continue operating the process. The companies that have invested in the best technologies will have a clear advantage over their competition and in some cases will force companies to reinvest in the latest techniques or discontinue product lines. I am convinced that at this stage the NEEC part of BATNEEC will become irrelevant. Companies that have protected their technology with patents will have an ideal opportunity to either sell environmental products or license new technologies to other companies. It is hoped that HMIP live up to their promises in these areas as many companies, like us, who have invested heavily in environmental technologies will be disappointed if they are not allowed to capitalise on their investments. If we fail in this area the heavy cost of IPC will only achieve its primary objective of improving our environment and not the spin off of commercial benefit to those companies who have been prepared to invest in their commitment to IPC.

2. ENVIROCATS

Envirocats have been developed as a joint collaborative project between York University and Contract Chemicals. The project has been funded jointly by Contract Chemicals and the Teaching Company Scheme. The catalysts are metallic salts supported on either clay or alumina and find their major use in Friedel-Craft and Oxidation chemistry. In Friedel-Craft chemistry we can replace the stoichiometric use of conventional reagents such as aluminium chloride with true catalytic amounts of Envirocats. This eliminates the large volumes of aqueous aluminium effluent created by conventional Friedel-Craft chemistry. Another disadvantage of conventional homogeneous catalysts is the production of unwanted side products and large amounts of polymeric material. Envirocats are non-toxic catalysts which can be easily reused and do not present any disposal problems. They are preferentially used in non-polar solvents with reactions very often requiring no solvent at all. Unwanted side products such as rearrangements or demethylations are eliminated or minimised and because of their heterogeneous nature the catalysts may be filtered off leaving no metallic residues in the product. Contract Chemicals has experience in the use of Envirocats in a variety of reactions.

Benzylation

p-Benzyl biphenyl has been produced by the benzylation of biphenyl using Envirocat EPZ10.

Nearly 100% conversion of the benzyl chloride is achieved giving a para/ortho ratio of greater than 3/1 and minimising the amount of unwanted polymeric bi-products. This is a better conversion than can be achieved using conventional catalysts. Moreover at the end of the reaction the catalyst is recovered by filtration and can then be reused. Up to ten reuses of the catalyst have been achieved in this type of reaction.

Benzoylation

4-Chloro benzophenone is produced by the reaction of 4-chloro benzoyl chloride with benzene in the presence of Envirocat EPZG.

Greater than 90% yield is achieved in this reaction without the use of a solvent. True catalytic amounts of Envirocat are used as opposed to the normal aluminium chloride reaction where greater than stoichiometric quantities are necessary. The catalyst is removed from the reaction by filtration and vast quantities of aqueous aluminium effluent are eliminated.

Oxidation

The oxidation catalyst EPAD is a non-toxic alumina based catalyst which is reusable. Because the use of solvent is usually eliminated, high space time yields are obtained and there is almost no effluent from the process. Typical reactions would be the oxidation of ethyl benzene to acetophenone and the oxidation of diphenyl methane to benzophenone. The oxidations are carried out in the liquid phase at about 140°C in the presence of a stream of air and with good agitation. The catalyst is again isolated by filtration and the products recovered by distillation. In this way the excess starting material and all of the catalyst may be recycled giving a process with almost no effluent.

The catalysts described are the first generation catalysts and already we have new catalysts in development. We are now producing catalysts with more activity and selectivity for use in less active systems. We see Envirocats as a way to promote environmentally friendly chemistry and hope that they will be considered as the best available technique in this area of chemistry.

3. PHYSICAL TECHNIQUES

It is the role of the development chemist to select the process route and optimise the efficiency of the process. Obviously the greater the yield the lower the waste products. Solvent selection plays an important role in the chemical design stage, the move being away from chlorinated and volatile solvents towards recoverable high boiling solvents or no solvents at all. I have already illustrated this with several of the Envirocat reactions. However, it is the collaboration between chemist and chemical engineer which can produce some major improvements in waste minimisation. Consider reactions of the type

$$R-OH + SOCl_2 \longrightarrow R-Cl + HCl + SO_2$$

where ROH can be an acid or an alcohol. The waste from the reaction is clearly the HCl and SO_2 gases which used to be absorbed in caustic soda to produce a disposable effluent. It is now clear that this is not the best environmental option. With the help of chemical engineers, it is possible to produce 2-stage scrubbing systems which can be used on small batch plants to produce reusable effluents. They operate by absorbing the HCl in water to produce commercial quality hydrochloric acid and absorbing the SO_2 in caustic soda to produce sodium sulphite solution. By operating this twin absorber system with low air flow reasonable quality products may be produced for further use. Although these systems have long been in use on large chemical processes, it is now necessary and desirable to use them on small batch processes. If we further consider the manufacture of sulphonates the sodium sulphite produced from the scrubber may be reused in the production of the finished product from the alkyl chloride.

$$R-OH + SOCl_2 \longrightarrow R-Cl + SO_2 + HCl$$

$$SO_2 + 2NaOH \longrightarrow Na_2SO_3 + H_2O$$

$$R-Cl + Na_2SO_3 \longrightarrow R-SO_3Na + NaCl$$

The recycling of vacuum pump water is now becoming a standard feature of our plants. In the past water used in liquid ring vacuum pumps was on a once through basis, the effluent generally being discarded to drain. This is no longer acceptable and recycle systems are now in operation on most plants and will certainly be fully introduced as IPC authorisations are approved. However, if you analyse the content of this effluent it is possible to reuse it. In our plant we use large tonnages of dimethylamine in the production of tertiary amines. Clearly our vacuum systems all have to be on recycle to eliminate the risk of dimethylamine escaping to the environment but by careful control of the recycle the amines can be reused in the process reducing the risk to the environment and reducing the unit cost of the finished product. As this process is aqueous the recycling of water is also important, not only minimising the loss of valuable product and raw material but also reducing sewerage costs.

Although we have a sophisticated scrubbing system for dimethylamine on our plant, it is our objective to recycle all the amines we use making the demand on the scrubber minimal. Abatement equipment should be used as a safeguard and not a means of absorbing the inefficiency of the plant.

One of the areas which is often neglected is the composition of the residues produced from processes, particularly from distillations. If we consider the production of benzyl biphenyl, which we looked at earlier, clearly there is a quantity of polymeric material produced from the process by subsequent reaction of product with more benzyl chloride. By treating these residues with a suitable catalyst it is possible to scramble them producing significant quantities of product which may be recovered. Development in the area of residue processing could have a significant impact on waste minimisation and simply by understanding the nature of our residues we may find ways of minimising their initial production. The use of advanced analytical techniques can help us identify the chemical species involved and collaboration with academic institutions can be a rewarding route to the solving of waste problems. In our case the residues from our amine process have been analysed by mass spectroscopy and the species identified have given us clues which have helped us alter our process conditions. This has resulted in an increase in yield and subsequent reduction in waste.

The separation and purification of products can result in significant waste production. Generally solvents are necessary for these separations with all the associated problems. Very often reactions produce ionic side products which have to be removed. A typical reaction could be

$$RCl + R_1NH_2 + NaOH \longrightarrow R-NH-R_1 + NaCl + H_2O$$

The salt impurity is usually removed by dissolving the product in a solvent, removing the salt by filtration and recovering the solvent by distillation leaving the product. The environmental implications of this process are not good. Solvent losses, energy costs and solvent damp salt cakes are just some of the problems. We have considered physical techniques as alternatives and believe they can be effectively utilised.

Electrodialysis

This is the technique by which ionic and non-ionic species are separated electrically across ion-selective membranes. This technique is under development for chemical manufacture and has been better developed for desalination projects. Trial work using this technique for the removal of ionic bi-products from fine chemicals has proved very encouraging with high purity products being produced. The disadvantages of this technique are high capital costs coupled with potential product loss by leakage across and around the membranes. It is also not suitable for materials containing particulate matter as the membranes are easily blocked and damaged and their replacement cost is high. If the industry shows more interest to stimulate development by the manufacturers, I believe the problems can be overcome and electrodialysis will become a valuable tool in fine chemical production. The environmental advantages of the technique are that the only potential pollutant is the usage of electricity, the salt separated can be of good purity giving ease of disposal or reuse.

Ion Exchange

Ion exchange already finds a variety of uses in the chemical industry but its use in waste minimisation has not been fully exploited. Ion exchange resins can be used for the removal of ionic impurities. Salts can be removed by using cation and anion exchange resins in series, the resins themselves being reactivated by treatment with simple acids and bases. Waste streams which contain trace quantities of acidic or basic products can be recovered using the appropriate resin and the product then eluted from the resin in a concentrated form with an acid or base. In this way valuable product can be recovered and a potential waste stream eliminated. Noxious waste streams can also be treated using ion exchange resins. In our case aqueous streams containing very small amounts of odorous amines can be treated with a cation exchange resin and the smell removed. The amines can then be removed from the resin with acid and are available for either reuse or disposal as a concentrated stream. The same technique could also be applied to awkward acidic bi-product streams. The resins are usually used in columns commercially and the systems can be automated to run continuously usually with pH control. The capital cost of the plant is not that high and although the resins are relatively expensive they can have a long life.

Solvent Extraction

People tend to view solvent extraction as an environmentally unacceptable method for product separation. This has been due to the use of volatile solvents and because of the considerable amount of solvent recycling necessary, leading to potential environmental pollution. Considerable engineering expertise is available in this area through the academic world, particularly Bradford University, who can provide pilot facilities to demonstrate all the latest techniques. High boiling solvents can be used in multi-stage systems to improve the effectiveness of extractions and because of the non-volatile nature of the solvent the potential waste is minimised. Interactive solvents are also available. High boiling acidic or basic solvents can be used to extract products forming liquid salts, this technique being very similar to a liquid ion exchange resin. By removing products from aqueous streams using solvent extraction we can minimise the high energy costs associated with large amounts of water distillation and show a great improvement in waste reduction over more traditional processes.

Reverse Osmosis

Reverse osmosis is the technique by which mixtures are separated as a result of the difference in molecular size. A membrane is selected with a specific pore size and the smaller molecules are forced through it under pressure. Traditionally the technique has been used to concentrate dilute solutions in the pharmaceutical industry where molecular size differences are large. It can also be used to separate simpler organic molecules. Although the separations may not be absolute where the molecular size difference is not great, the concentration increases that can be achieved are significant and can decrease the work load on a more traditional separation. The concentration of dilute aqueous solutions of a medium size amine molecule by this method has been demonstrated to be an effective method of recovering the product. Where molecular weight differences are large, perhaps in the case of the polymer industry, the technique can be used with larger pore size membranes. This is then called Ultrafiltration.

I am not suggesting that physical techniques provide the answer for all your problems, but by being aware of their availability you may try them on your processes and if they are successful, improvements can be made in environmental performance.

It is only by trying different techniques and pursuing new technologies that the goal of experimental improvement can be achieved.

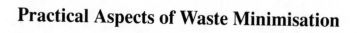

Practical Aspects of Waste Minimisation

The Interface between the Chemist and the Chemical Engineer as a Source of Waste

Robin Smith[1] and Eric A. Petela[2]

[1]CENTRE FOR PROCESS INTEGRATION, DEPARTMENT OF CHEMICAL ENGINEERING, UMIST, PO BOX 88, MANCHESTER M60 IQD, UK

[2]LINNHOFF MARCH LTD, KNUTSFORD WA16 OPL, UK

1   INTRODUCTION

Figure 1, the "onion diagram", depicts the hierarchy intrinsic to chemical processes. We start with the reactor in the core of the onion. Whatever design is chosen for the reactor dictates the separation problem and process recycles. Once these two layers are fixed we know the material and energy balance and can design the heat recovery system, the third layer in Figure 1. After the economic possibilities for heat recovery have been exhausted we add utilities (furnaces, steam etc), the fourth layer in Figure 1.

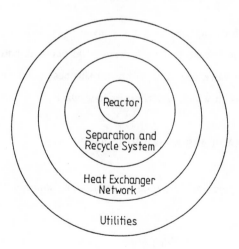

Figure 1   The Onion Diagram

Referring to the onion diagram in Figure 1, we can identify two classes of waste from chemical processes[1]:

(a)   The two inner layers of the onion (the reaction and separation/recycle systems) produce *process waste*. The process waste is waste byproducts, purges, etc.

(b)   The outer layer of the onion (the utility system) produces *utility waste*. The utility waste is products of fuel combustion, waste from boiler feedwater treatment, boiler blowdown, etc.   Clearly the design of the utility system is closely tied to the design of the heat exchanger network.   Hence in practice, we should consider the two outer layers as being the source of utility waste.

Waste generated in the reactor can produce major environmental problems. Even worse, since the reactor is at the heart of the process, any waste created here generates extra waste throughout the design. The extra waste created in the reactor needs extra separation. This tends to reduce the efficiency of the separation system and increases the energy demand.

There is an obvious priority to first consider the reactor system when looking to minimise waste. This is where the chemist and the chemical engineer need to work together to obtain the best solution. Let us now consider the ways in which reactors produce waste.

2   REACTORS AS SOURCES OF WASTE

Under normal operation there are five major sources of waste from reactors[2].

-     Low conversion if recycling is difficult
      If it is not possible to recycle unreacted feed material back to the reactor inlet, then low conversion in the reactor will lead to waste

of that unreacted feed.

- Waste byproducts from the primary reaction
  The primary reaction can produce waste byproducts, eg:

  FEED 1 + FEED 2 ⎯⎯⎯> PRODUCT + WASTE BYPRODUCT

- Waste byproducts from secondary reactions
  Secondary reactions can produce waste byproducts, eg:

  FEED 1 + FEED 2 ⎯⎯⎯> PRODUCT
  PRODUCT ⎯⎯⎯> WASTE BYPRODUCT

- Feed impurities undergo reaction
  Impurities in the feed materials can react to produce additional
  waste byproducts.

- Catalyst waste
  Catalyst is either degraded and requires changing or is lost and
  cannot be recycled.

Let us now take a brief look at each of these sources of reactor waste.

3   GETTING THE RIGHT REACTOR

Reactor Type

To minimise waste generation in the reactor we should first make sure that
the correct reactor type has been chosen. There is an enormous variety of
reactor types used in the process industries. When trying to minimise
waste generation, our approach is first to select the ideal characteristics
we would require. These ideal characteristics are well defined by reactor
design theory. We then choose a practical reactor to approach as closely

as possible the ideal characteristics.

Reactor Conversion

If recycling is not possible we should, in general, try to achieve a high conversion in the reactor. If the reaction system involves a single, irreversible reaction then a higher conversion can be achieved using longer residence time, higher temperature or higher pressure. The problem of forcing a high conversion with single reactions becomes acute if the difficulty of separation and recycle coincides with the reaction being reversible. There is a maximum conversion, the equilibrium conversion, which we cannot exceed even with long residence time.

- To force high conversion in equilibrium reactions adjust the feed ratio, eg use excess of one reactant.

- Change the inerts concentration, eg add inert material if reaction leads to increased number of moles.

- Change the temperature, eg raise temperature for endothermic reaction.

- Change the pressure, eg increase pressure if reaction leads to a decrease in the number of moles.

- Carry out intermediate separation of product as the reaction proceeds.

Waste Byproducts from the Primary Reaction

If the reaction is of the form:

FEED 1 + FEED 2 ⟶ PRODUCT + WASTE BYPRODUCT

Then we can usually only avoid the formation of waste by using a different reaction path. This will likely involve a change in feedstock, different reaction chemistry and possibly a totally different process.

Waste Byproducts from Secondary Reactions

A typical example of this situation is the following reaction system:

FEED 1 + FEED 2 ———> PRODUCT

PRODUCT ———> WASTE BYPRODUCT

However, there are many classes of multiple reactions. The action necessary to minimise reactor waste depends on the class of the reaction system[2].

Feed Impurities Undergo Reaction

If one or more feeds to the reactor contains an impurity, this impurity can become waste or even worse could react to produce waste byproduct. Clearly, the best solution to feed impurities is feed purification. However, the decision as to whether to purify the feed or purify the product should be taken bearing mind all the associated costs, ie raw materials, feed purification and waste treatment.

Catalyst Waste

We can reduce waste from catalysts by using heterogeneous catalysts rather than homogeneous catalysts, which are often difficult to separate and recycle. In addition, we should look to prolong catalyst life which can be done by avoiding local extreme conditions. Actions we can take to help safeguard our catalysts include:

- better flow distribution

- better heat transfer
- introduction of catalyst 'diluent'
- better instrumentation and control

4   THE INTERFACE BETWEEN THE CHEMIST AND THE CHEMICAL ENGINEER

Large scale processes in the processes industries such as those typically found in the petrochemical industry usually feature reactors which are well optimised. Such process are typified by:

- Continuous operation
- Large throughput
- Unchanging products
- Simple chemistry (relatively)
- Mature process technology

The optimization leads to a wide variety of reactor configurations (tubular reactors, fluidised beds, etc) tailored to suit exactly the requirements of the process. The optimization also usually recognises the contribution the reactor makes to waste on the process. Unfortunately, whilst the optimization will no doubt have featured the cost of the raw materials lost to waste byproducts, it has not been common practice in the past to include in the optimization those waste disposal and effluent treatment costs which arise from the formation of the waste byproducts. If such costs are included it is possible that significant changes could occur in the trade-offs to indicate a lower economic level of waste formation. However, the scope to reduce waste formation in such processes is often limited unless, for example, a breakthrough is made to find a new catalyst with improved selectivity.

On the whole, these large continuous plants are good examples of the chemist (with his knowledge of the chemistry of the process) working with the chemical engineer (with his knowledge of process systems) to achieve an optimum overall design and one which typically address waste at each stage of design and operation.

All of this tends to be in stark contrast to small scale speciality chemical processes. Here the chemistry tends to be complex. Perhaps because of this, more often than not the reactor is a direct scale-up from the laboratory. Since the chemist carrying out the laboratory investigation will almost invariably work in batch mode, the production plant also becomes batch. Whilst batch reactors can have significant advantages in some circumstances, e.g. if the process is multi-product, etc, they also have many disadvantages:

- Difficulty in maintaining optimum reactor conditions
- Start-up and shutdown losses
- Use of excess material in processes not designed for recycle
- Inefficiencies in the utility system caused by varying conditions.

In these processes the same basic design of stirred tank tends to be used irrespective of the reaction system. If unreacted feed material can readily be separated and recycled then a low conversion in the reactor can often reduce waste formation. Yet, small scale speciality chemical processes tend to operate at as high a conversion as possible; probably because this is the mode in which chemists will work in the laboratory. Conditions of reaction carried out in the laboratory are often limited by the limitations of laboratory equipment. Yet the chemical engineer is not limited in this way when designing the full scale process. Extraneous materials added to the process by the chemist in the laboratory, for example to promote precipitation, tend not to be a problem in that context. Yet, those same extraneous materials in the full scale process can lead to major sources of waste. Unless such points are understood between the chemist and the chemical engineer then unnecessary waste formation can result. The chemist is the expert in the reaction chemistry, the options possible, reaction conditions, etc. The chemical engineer needs to understand from the chemist the overall design area in which he is allowed to work so that he can use his knowledge and ability to achieve a good design.

5   FINAL COMMENTS

The theory of reactor design is well established. Given sufficient information about the reaction chemistry we can direct the theory to minimise the waste formation. Whether or not we are in command of all the information we would like, the reaction system will fall into one family or another. Our approach must therefore be to establish which family a reaction system belongs to by a minimum of laboratory work and then direct the theory to minimise the waste production.

At this point it is important not to allow artificial constraints from the laboratory to impose a design on the plant. Artificial constraints can lead to unnecessary waste formation. Left to their own devices a chemist would probably never have a continuous process and the chemical engineer would never have a batch process. The interface between the two should ensure that the most appropriate design is chosen.

6   REFERENCES

1.   R Smith and E A Petela, The Chemical Engineer, 1991, No 506
        31 October, p24.

2.   R Smith and E A Petela, The Chemical Engineer, 1991, No 509/510
        12 December, p17.

Waste Minimisation – The Role of Process Development

I. G. Laing

FORMERLY CIBA UK GROUP, HULLEY ROAD, MACCLESFIELD, CHESHIRE
SK10 2NX, UK

1. GENERAL

Process development is a long-established traditic ι. in the chemical industry, necessary for scale-up from laboratory to plant, or, with established plant processes, for making economic improvements or improving the quality of the end-product. It is, however, also a powerful tool in minimisation of waste.

Generation of waste represents a depletion of resources, mostly non-renewable. The basis of an effective approach to the minimisation of waste and hence minimisation of adverse environmental impact is embodied in a four-step hierarchy:-

1. Reduce pollution at source.
2. Recycle or re-use waste streams arising despite efforts to reduce pollution at source, i.e. maximise recycling/re-use options to transform to a form of value to a third party if not to the waste generator.
3. Treatment of all waste streams that cannot be avoided, recycled or re-used by the method having lowest environmental impact, i.e. render 'safe'.
4. Disposal of ultimate residual waste safely.

From the development standpoint, the two most important techniques at our disposal are (1.) source reduction and (2.) recycling/re-use. Techniques concerned with (1.) source reduction, are summarised in Figure 1. Good management practices, although important, are not within the scope of this paper and will not be considered further. Recycling/re-use (2.) covers activities on- or off-site, use/re-use in the processes from which the waste originated, or in other processes, or other types of reclamation/resource recovery. It is, however, reduction in consumption of materials, in particular, changes in materials used and process conditions/ technology, on which most emphasis will be placed.

Figure 1

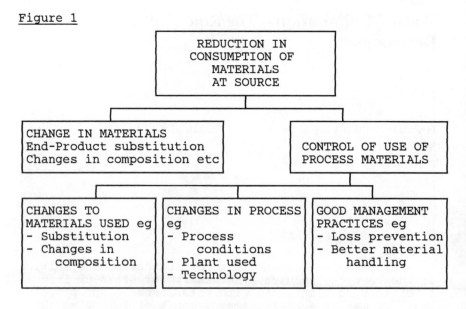

2.   SOURCES OF WASTE IN CHEMICAL PROCESSES

The chemical industry differs in one essential aspect from other equally technical industrial sectors; the uniqueness of chemical reactions. Transformation of raw materials into downstream products is the result of chemical reactions, which rarely run 100% in one direction. There are invariably traces of unreacted raw materials present in the final reaction mixture and also the products of side reactions arising from the main reactants. Such by-products are in most cases undesired, must be separated from the desired end-product and are rarely re-used or recycled. Furthermore, a wide variety of auxiliary chemicals are added to chemical reactions, e.g. catalysts, solvents, and these too are present in the final reaction mixture and may in turn contribute to the formation of by-products. Impurities present in any of the process raw materials will also remain in the final reaction mixture or form by-products. This situation, described as the 'anatomy of a chemical reaction', is summarised in Figure 2.

3.   THE PROCESS BALANCE

Chemical processes are governed by laws of nature and to fully understand the process, information on mass and energy balance is essential. The Process Balance (Figure 3) is a description of the process summarising all input and output streams, likewise the energy balance. It provides the basis for identification and quantification of waste reduction opportunities and

Figure 2

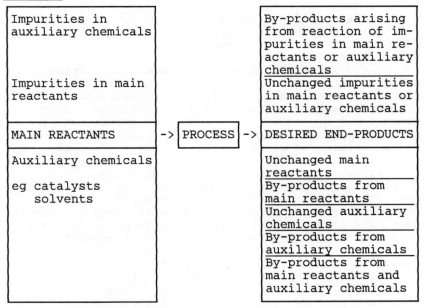

demonstrates what happens to all materials fed into the chemical process.

Figure 3

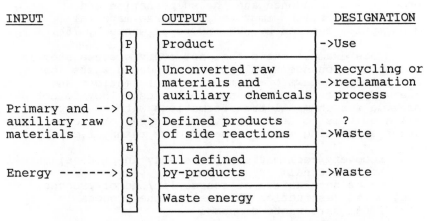

To make progress in the reduction of chemically generated waste it is essential to know not only the analytical composition of all feedstocks, but to have a thorough knowledge of the properties and constitution of all exit-stream components. This is reflected in the four-stage action programme that should be followed when embarking on a development study.

1. Identification, characterisation and quantification of waste streams.
2. Investigation of waste reduction possibilities.
3. Setting targets/priorities.
4. Allocation of resources.

The case of 2-chloro-5-nitrotoluene-4-sulphonic acid can be used to explain the origin of waste in a chemical process and illustrate how information useful for the reduction of waste in such a process can be effectively applied. This product was traditionally produced by a single-pot procedure resulting in about 60% conversion of raw materials to the desired product and 40% to waste. The so-called one-pot process starting from toluene, in which three chemical steps, sulphonation, chlorination and nitration, were carried out consecutively in one reactor, was almost certainly the result of process development work aimed at reducing cost. The analysis of the various streams in the mass balance reveals an extended system of side-reactions (Figure 4.) with no fewer than eleven by-products. Comparing the desired reaction pathway with the mixture produced by the applicable laws governing chemical transformations, shows too that if not all undesired products originate in the first and third steps, whilst the second step, the chlorination stage, does not contribute as much to the growth of undesired by-products. This knowledge derived from the mass balance then provides the base for improvements of the existing high-waste process to an improved low-waste process. Without going into details the single pot process was abandoned and the sulphonation and chlorination stages improved, with raw material consumption reduced by 50% and total waste by 75%.

This example shows that not every process step is amenable to change resulting in a reduced waste load. The universal laws of chemical transformations and thermodynamics limit the possibilities. They do not provide for chemical transformations with no waste at all and it is to deal with these natural obstacles that other supporting measures have to be taken, i.e.

a) recovery/reclamation (preferably in-process, but if not, external)
b) waste destruction at source (in/end of process)
c) waste destruction external to the process
d) final safe waste disposal

4. CHEMICAL PROCESS DEVELOPMENT : OPTIONS AND EXAMPLES

There are between 100 or 150 different chemical reactions of industrial importance. This relatively small number of reaction types is used in thousands of

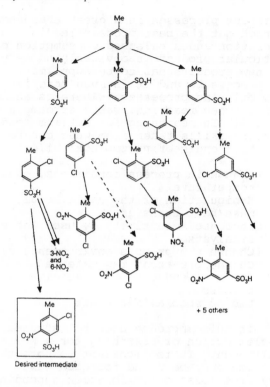

Figure 4.

Production of 2-chloro-5-nitrotoluene-4-sulphonic Acid

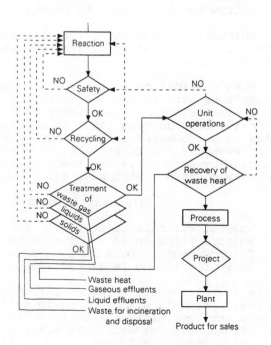

Figure 5.

Integrated process development

different processes in a great abundance of variations.
To pick out the best process leading to minimum waste
generation and a balanced consumption of resources for a
particular product is obviously no easy task. Research
for new products has to be supplemented by research for
new processes, and it is even more important to evaluate
anew existing processes. There is an intimate relationship between reaction, waste streams and final plant
process. In searching for optimum conditions one applies the skills taken from what can be described as the
'reaction engineering toolbox', i.e.

- Change in process conditions, e.g. pressure,
  temperature
- Manipulation of the equilibrium, kinetics,
  mass/heat transfer
- Increase selectivity by use, for example, of
  catalysts
- Change the process environment, e.g. solvent
- Change/pre-treat raw material
- Internal and external recycling and reclamation
  process
- Use of specialised plant

If this approach does not succeed, the even more
complex option of searching for a new chemical path is
still open. It is important to note that a tailor-made
solution must be found for each case, which might even
vary from location to location depending on external
circumstances. R&D have to exploit the possibilities of
alternative chemistry but the possibilities are limited.

Sulphonation and nitration are extremely important
reactions and the following examples illustrate the
challenge facing R&D.

When considering routes to sulphonated aromatic
hydrocarbons such as benzene or naphthalene, there is no
useful alternative to use of sulphuric acid or one of
its derivatives as main reaction component. Current
routes to sulphonated aromatic products in general are
not necessarily interchangeable and have to be evaluated
on a case-by-case basis for their suitability to produce
the desired product with high efficiencies and low
waste.

Sulphonation Methods

1.   Sulphur trioxide in sulphuric acid H_2SO_4/SO_3
2.   Sulphur trioxide in a chlorinated hydrocarbon
3.   Chlorosulphonic acid in a chlorinated hydrocarbon
4.   Sulphur trioxide in liquid sulphur dioxide

Of the four alternative routes for sulphonation, (1) is
the most wasteful, using sulphuric acid itself as
solvent for the reaction, but is still commonly used.
It is frequently impossible to achieve high recycle
rates of sulphuric acid and, when feasible, these are

very energy consuming. Routes (2) and (3) do not
generate sulphuric acid wastes but the use of
chlorinated hydrocarbon solvents is the price to be paid
for this advantage. These solvents exhibit high recycle
potential at very moderate energy consumption and
produce residues for incineration that are easy to
handle. Route (3) produces, in addition, hydrochloric
acid as a by-product, which can be 100% recycled in the
process itself as will be shown in a working example
later. Route (4) at present is only a laboratory
curiosity with the disadvantage of using pressure and a
solvent that is not very easy to handle, namely liquid
sulphur dioxide. It would require heavy investment in
plant, e.g. it is only likely to be justifiable for
large production volumes.

An example of nitration is the conversion of
anthraquinone to 1-nitroanthraquinone. There are three
options:-

Nitration of Anthraquinone

Nitration Agent	Yield (% Theory)
A) Nitric acid in concentrated sulphuric medium	60
B) 100% Nitric acid	75
C) Nitric acid in hydrofluoric acid medium	>90

The efficiencies of each indicate how much effort has to
be invested to separate the good product from the waste
in the reaction mixture. Lower yields result directly
in higher energy consumption and additional waste processing. Route (A) carries a very high energy demand
for processing and recycle of the waste acids. Recycling
of the spent nitric acid in route (B) is comparatively
energy efficient. The same applies to route (C). So, if
we compare overall energy and waste processing demand
per unit product produced, process (C) is highly favour-
ed over (B) which in turn is much better than (A). It
is process (A) that is still used on industrial scale
today. Process (B) failed on industrial scale-up and
Process (C) is still a laboratory and patent curiosity.
The most efficient process may well combine working in a
highly aggressive environment and using pressure.

The reduction of nitro aromatics to arylamines is an
important basic reaction widely used in the specialised
organics sector. Traditionally iron or zinc were used as
reducing agents, giving rise to the corresponding metal
oxides as by-products and potential waste. Application
of catalytic hydrogenation has been widely used in the
last 20/30 years extremely successfully, giving a much
cleaner and more efficient reaction (fewer organic by-
products) and without any waste streams derived from the
reducing agent. This change required capital investment

and a shift from traditional reaction vessels to loop reactors.

When there is an opportunity to build a new plant, the search for improved environmental performance must be a priority even though it may not be the main reason for the project. A good example is provided by a new plant built for a range of speciality organic products at the Clayton Aniline Company plant in Manchester. The main initial objective was an increase in capacity, but other major targets were to reduce the environmental impact, lower costs, improve material handling, minimise the range of solvents used, etc. All the processes were re-developed and achievements were substantial - necessary increased capacity, reduction in costs by 27%, 90% reduction of total organic carbon in the aqueous waste stream, elimination of use of a variety of chlorinated solvents and use of only two non-chlorinated solvents both recovered and recycled. Such organic contaminants as persisted in the aqueous waste stream are biodegradable and there was also a reduction of inorganics (principally sodium sulphate) in the waste water by 67%.

At the same plant another good example can be provided in reduction of copper waste in a series of copper-catalysed reactions of Bromaminic Acid (1-amino-4-bromo-anthraquinone-2-sulphonic acid) with arylsulphonamides. By modifying the physical form and hence efficiency of the copper compound used as catalyst, considerably less catalyst was necessary and the copper load in the process effluent and the level of copper found in the end-product, were both reduced by circa 97%.

5.   INTEGRATED PROCESS DEVELOPMENT

Here the aim is basically to modify the processes to minimise the waste streams in the first place, but for the wastes that are unavoidable, to integrate recovery or treatment/destruction into the process. Integrated process development was identified as the way forward by Ciba-Geigy in 1978. The uniqueness of this procedure is that already during the development stage of a production process, type and quality of starting materials, optimal evaluation of re-cycling options and improved treatment or elimination of process waste streams must be considered as a whole. This requires, in particular, that for each production process, optimum chemical and physical conditions must be established, first in the laboratory and then at pilot stage. Process balances must be established and technologies for the optimal treatment of waste streams developed. The search for the best process, which will lead to the desired product with a minimal generation of wastes and an optimal use of resources, is a challenge for all the professionals involved, e.g. chemists, chemical engineers and engineers. This is a teamwork exercise from the outset. In Integrated Process Development, waste treatment is moved from 'end-of-pipe' to 'end-of-process' or even 'in-process'.

The team must consider recycling in detail at the development stage. Recycling can be described as the 'operative element' of resource conservation, and in the context of waste minimisation, recycling includes all measures which allow a re-use or putting a value on (recovering) process residues, either within the original process itself or elsewhere. Whether or not this is possible, and sensible in every case, depends on a number of factors, e.g.

- Quality and properties of the material
- Options for the use of the material
- Availability of similar or competitive substances on the market, and their costs
- Availability of a specific technology or possibilities for its development, with the purpose of obtaining a substance of the desired quality
- Assessment (technical, ecological, economical) of the impact of the non-recovery of the substance, and in that case
- Assessment of long term risks and liabilities
- Logistical constraints
- Questions of operational safety and health in connection with recycling
- The time factor

Calculation of the process balance invariably shows that recycling processes require a relatively large input of energy. In addition, recovery processes invariably in turn become a new source of waste themselves. Frequently the results of an evaluation may show that recycling is not an economical option, but this may only be because todays energy costs even now do not adequately represent the true value of scarce and non-renewable resources. Under certain conditions the recovery and use of the heating value of a material can be a better ecological solution than its re-use or recovery.

Where recovery isn't possible the development team will also consider the best form of waste-stream treatment to incorporate, e.g. in the case of an aqueous waste stream

- biological treatment
- chemical oxidation/adsorption
- wet air oxidation
- combustion

In working up an Integration Process Development Project, a decision tree approach is followed (Figure 5). This approach illustrates the relationship between production process, plant and waste streams, with part-

icular emphasis on health and safety, which forms the
first 'challenge'. Integrated process development
allows for a dynamic search for optimal processing conditions by iterative use of the established tools of
chemists and engineers. There are no standard recipes
for the solution of the acknowledged problems of waste
disposal. Each case must be investigated individually
and requires a tailor-made solution, which is then specifically adapted to each production plant because of
specific differences in technical conditions. The
described approach for environmentally sound chemical
production applies to individual process stages as well
as to complete, complex, multi-stage processes. The
achieved improvements will be reflected in the environmental balance of a production facility.

Examples of Integrated Process Development

a)   Transformation of an organic acid into an amide
     used as a Pharmaceutical intermediate

A conventional method for the transformation of an
organic acid into its amide is via the corresponding acid chloride which in turn is obtained by the
reaction of phosphorous trichloride with the organic acid. The reagent as well as the acid chloride
produced are both very reactive, therefore particular precautions have to be taken for the synthesis.
Undesirable side-products include organic wastes,
phosphoric acid and a large amount of salts.

The process was reviewed by R&D and a new reaction
path was found, which eliminated the critical
stage. As a result, a new synthesis for the production of the amide was developed, which in turn
resulted in the building of a new production plant.
The new concept produced no effluent stream. The
only output streams of materials which leave the
plant besides the desired amide, are a pure, recyclable acetic acid as well as an easily combustible
distillation residue, which contributes to heat
recovery (see Figure 6.). Such a fundamental review
of a process stage required the resources of one to
two process chemists/engineers for approximately
one year.

(b)   Dinitrostilbene disulphonic acid ('DNS')

'DNS' is an important intermediate for dyestuffs
and optical brighteners. The starting material,
4-nitrotoluene-2-sulphonic acid (PNTSA), is
oxidised to form 'DNS'. There are three alternative processes:-

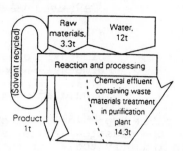

Figure 6.

Pharmaceutical intermediate production

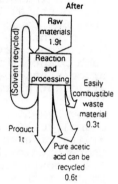

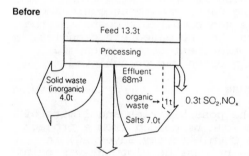

Figure 7.

H–Acid production

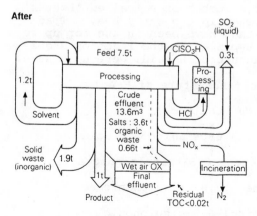

                    Oxidation of PNTSA to DNS
Oxidant   Reaction Medium   Yield

NaOCl   Water   60-75%

Oxygen   Dimethyl formamide or
                       Dimethyl sulphoxide   > 90%

Oxygen   Liquid ammonia   > 90%

Using sodium hypochlorite as oxidant in aqueous
medium, the classical route, still the main process
used in industry today, it took about eighty years
of industrial R&D to increase process efficiency
from 60% to 75%. The next step was the development
of solvent-based processes which resulted in the
much higher process efficiencies (> 90%) and one
such plant is in operation. However, the solvent
process has a potential rival in an equally
efficient process using liquid ammonia as the
'solvent' or reaction medium involving full
recovery which, if implemented, would be a prime
example of a technologically more complex process
giving rise to the least waste. R&D costs for
these studies have been in excess of 5 million $s.
Plant costs including waste treatment could amount
to about three times the comparable costs for the
old process. Due to the increased efficiencies,
and decreased waste treatment costs, economic
parity with the old process can be achieved.

(c) The 'Letter' Acid case

More insight into the possibilities and limits of
integrated solutions can be gained from a project
realised at Schelde Chemie (today Bayer AG) in
Germany. Whilst the process development was being
carried out, Schelde was a joint venture between
Bayer and Ciba-Geigy. This world-scale plant for
the manufacture of a number of so-called 'Letter'
acids was commissioned in the early 80s.   Traditional processes that have been in use for up to
eighty years, and typified by the reaction steps
involved in manufacture of H-Acid, were up-graded.

            Steps in the manufacture of H-Acid

                    Naphthalene
                        ↓
        Naphthalene-1,3,6-trisulphonic acid
                        ↓
    1-Nitronaphthalene-3,6,8-trisulphonic acid
                        ↓
      1-Naphthylamine-3,6,8-trisulphonic acid
                        ↓
      1-Amino-8-naphthol-6,8-disulphonic acid

The result was low waste processes with integrated recycling and waste treatment, and the process balances of these processes showing the situation before and after development work tell the story (Figure 7).

In summary, the following benefits were achieved :-

- Raw material consumption reduced from 13.3t to 7.5t per ton of end-product - a reduction of 44%
- Solvents recycled in process (no solvent losses or emissions)
- Hydrochloric acid (gas) converted back to Chlorosulphonic acid and recycled in process
- Sulphur dioxide (gas) purified, liquified and sold for external recycling
- All wastes reduced by about 50%
- Aqueous effluent reduced by about 80%
- Organic load in aqueous effluent reduced by more than 97% through integration of wet air oxidation
- Waste gas reduced to nil through inclusion of waste air incineration

The work required the equivalent of 60 man-years effort for chemists, chemical engineers and engineers. Obviously there is a cost associated with this type of progress. In todays money development costs for this integrated solution would exceed $50 million, and the cost of the new plant would be high. It has to be stressed that large production volumes, similarities in products and the possibility of building a new plant at a new site facilitated a near perfect solution.

6.   SUMMARY

There are no ready made recipes for waste avoidance with chemical processes. Each process has to be developed on its own merits. Conservation of resources is the governing principle, e.g. a balanced consumption of raw materials and energy coupled with secure waste facilities. This includes planned recycling of by-products and wastes as well as recycling of wastes for energy.

Excellence in low waste processing is achievable. Five steps to excellence in low waste processing can be identified:-

- Reduction of resources consumed
- Recovery and internal/external recycling
- Low waste processes/technology
- Process integrated waste destruction/disposal
- New chemistry

    Waste minimisation ultimately is part of a strategy for sustainable development and therefore an investment in the future, i.e. the design and development of products and processes in such a way that all aspects of (i) resource conservation, (ii) safety, health and hygiene, and (iii) environmental protection are taken into account.

Advances in Chemical Recycling for Plastics and Elastomers

Viktor Williams

MANAGER, WASTE MANAGEMENT PROGRAMS, EUROPE, DUPONT DE
NEMOURS INTERNATIONAL SA, PO BOX 50, 2 CHEMIN DU PAVILLON,
CH-1218 GRAND-SACONNEX, GENEVA, SWITZERLAND

Waste, and specifically polymer waste, is a very real -
and growing - issue in Europe.

The quantity of polymer waste is relatively small
compared to the total solid waste generated. Automotive
waste is far from being the biggest contributor to the
solid waste stream. However, the automotive industry is
under intense scrutiny over the polymer waste issue.

Indeed, it is the second industry after packaging on the
legislators' agenda. Draft regulations already published
by the German government in August 1992 put the
responsibility for managing waste from used cars firmly
on the automotive industry.

For elastomers used in the automotive industry, the
German government has stipulated that 20% must be re-used
or recycled by 1996.

This presents the plastic and rubber industries, from
material suppliers to end-users, with a considerable
challenge.

The average life cycle of a car is 10 years. But when
its useful life is over, the car has to be disposed of in
an environmentally acceptable way. Last year, some 13
million cars were scrapped in Western Europe.

Ninety-five percent of the metal waste from cars is
recycled. Yet 20 - 25% of the car, by weight, is non-
metallic: glass, textiles, plastics - and rubber.
Proportionally to the car's weight, polymers' share is
growing - from 4 to 6% in the mid-1970s to between 10
and 16% now.

This means that plastic and rubber waste from cars is
projected to increase from 0.8 million tons in 1990 to
1.9 million tons - more than double - by the year 2000.
How do we dispose of this automotive waste ?

* DuPont's registered trademark

Let me stress that whatever the material, it is essential that the collection and dismantling logistics of the waste retrieval system be in place first. This has direct bearing on the effectiveness of the disposal methods - including recycling, whether they be mechanical or chemical.

There is a German scheme which may become the forerunner for a European approach to this issue.

The scheme requires the last owner of the car to bring it to a licensed vehicle dismantler who issues a certificate of acceptance, which the owner can then use to cancel the existing tax and insurance contracts.

The vehicle dismantler first drains the car of all fluids - fuel, oil, grease, anti-freeze. He then removes spare parts that can be re-used, before handing over the other parts - plastic, glass, rubber - to the respective materials industries to provide help for recycling and disposal.

The car body - dry and with a high steel content - is given to the shredder. The steel goes for recycling. The remaining waste or fluff is currently either incinerated to energy or landfilled.

As far as the plastic and rubber parts are concerned, the economics are crucial in this scheme. The bigger parts are dismantled first. They are easier to disassemble and they usually consist of a single polymer. The smaller the parts, the more difficult to dismantle and the more costly.

The value curve of the material increases sharply at the beginning. But the fact that these parts are smaller and more contaminated means that they don't actually add very much to the total value.

So as you can see, we reach a break-even point beyond which the economics are unattractive, and an optimum point where you have the best overall economics. This is represented by a total of 5 to 7 parts.

These parts will most likely not include rubber parts except for the tyres, since rubber parts will be highly contaminated by oil and other fluids and hardly suitable for subsequent mechanical recovery.

The point I am illustrating is that it is frankly unproductive to try and dismantle all the polymer parts in the car. Economically, it makes sense to concentrate on the 5 to 7 parts I mentioned, which represent 20 - 30% of the total plastic weight in the car.

A bumper, for example, can be dismantled in relatively little time, costs little to do so and brings a relatively high value return.

Now, you can make this approach more attractive by easing the dismantling operation, through changes in the design of the part; using fewer plastic types or a coding or marking system to identify plastic types.

These are the findings of a PRAVDA pilot project which aims to work out the logistics and costs of automotive dismantling as a basis for recommendations for legislation.

PRAVDA is a German consortium of car manufacturers and polymer suppliers to the automotive industry which has set up automotive dismantling facilities in six German towns. PRAVDA is organized into the different groups of plastics and DuPont is a member of the nylon group.

I must point out that whatever the changes we decide on today in the design of parts or the selection of resins, we will only see improvements in 12 - 15 years' time because of the long life cycle of today's cars.

DuPont also participates in a German rubber industry initiative called Gesellschaft fuer Altgummi Verwertung System or "GAVS". GAVS has to deal with the 1.5 million tons of German rubber scrap generated annually - 2/3 of it being tyres - and its sensible recovery. I will come back to GAVS' activities later.

Now, many believe, legislators included, that mechanical recycling is the "magic cure" for all our waste ills. But I put it to you that mechanical recycling is not a long-term solution to the question of automotive or other kinds of polymer waste. Because with mechanical recycling, you are simply postponing the issue: you have a lower quality recyclate; the economics are not always attractive and new end-users are difficult to find.

The true solution for the future is chemical recycling. In the olden days, alchemists were employed by kings to transform low value objects into gold. Today's alchemy, chemical recycling, is a state-of-the-art, chemical procedure to break down the molecules of low value plastic and rubber scrap to get back to high value oil, oligomers or the original monomers; you actually end up with the virgin substances used to make polymers in the first place.

The big difference is that while old-style alchemy remained a pipedream, intensive research and development on the part of chemical companies like DuPont over the past years means that today, several chemical recycling technologies are available and working.

These include glycolysis, methanolysis, pyrolysis, gasification and hydrogenation, a technology for transforming plastics back into their pure state, which is oil, which can then be used as a feedstock for refining operations as though it were virgin petroleum oil.

Let's first discuss resin-specific technologies. For the core automotive industry thermoplastics produced by DuPont - ZYTEL* nylon 66 resin, DELRIN* acetal resin and RYNITE* polyester resin - our company has produced specific technologies that can be used to make oligomers or, ultimately, the original monomers.

DuPont recently announced that it has a pilot plant in the U.S.A. for a proprietary chemical process that can handle a mixture of nylon 66 and nylon 6 together - both fabricated nylon products and fibers. This obviously simplifies the logistics of collection and sorting. We have plans for European facilities and are developing standard nylon grades that have a recycled content, which will eventually complement our existing product line.

Another chemical recycling technology, this time for our acetal homopolymer resins, also looks promising. The problem here is that the acetal market tends to be rather limited, so that collection and sorting of these small parts is even more complex than for nylons.

For the chemical recycling of polyesters, we already have a DuPont glycolysis plant in Uentrop, Germany, and another in the U.S.A. The recyclate is made into quality fibrefill for anoraks, sleeping bags and other items.

And we have a patented methanolysis technology for polyester, to take it back to ethylene glycol and DMT - the original monomers. The beauty of this process is that the separation of impurities in the feedstock is much easier than any other depolymerization technology for PET. As a result, the feedstock can be as varied as you can imagine - from PET bottles to PVCD-coated film.

The best example I can offer of the advantages of chemical recycling for resin specific polymers, using our DuPont technology, concerns the tall ship, the HMS Rose, whose 17 beautiful, top quality sails were originally - in a former life ! - a mixture of car fenders and soda bottles. That's right, the sails are made of 100%

recycled polyester; no virgin material went into the sailcloth.

Chrysler car fenders made from DuPont polyester composite were repolymerized using depolymerization to get back to the basic molecular building blocks. This was then repolymerized into a resin compatible with pellets mechanically recycled from 126,000 plastic soda bottles. The pellets were spun into yarn and the yarn made into sailcloth.

Doesn't that remind you of alchemy - low value to high value ?!

Now on to mixed polymers. For elastomers, the main chemical recycling technologies that we're working with are: gasification, degradative extrusion, hydrogenation and catalytic extraction process (CEP). Let's now look at these for a few minutes.

For gasification, the feedstock can contain halogens, but must be free of metal and glass. The technology really consists of treating these mixed polymers at high pressure of 15 - 18 bars and at extremely high temperatures of 1,200 - 1,500 °C. The output is syngas consisting mainly of carbon monoxide and hydrogen.

Degradative extrusion can accept halogen-containing polymers, but we don't yet know the percentage. The feedstock must be metal and glass-free. The technology consists of a double extrusion, first at 250 °C where the halogen is removed and neutralized, then a second extrusion at 450 °C. The output can be either a light oil mix, gasoline, gas, fuel, or refinery feedstock.

The U.S. company Molten Metal Technology uses patented catalytic extraction processing (CEP) to dissolve mixed polymers into a molten metal bath operating at 1,650 °C, which separates the waste into its constituent elements. Polymers containing a high degree of both halogen and metal can be accepted. The output is separated into 2 product streams; metals and gases suitable for subsequent synthesis in the petrochemical industry.

These technologies usually need huge investments to the tune of several hundred million Deutsche Marks in order to justify economy of scale. The exceptions are degradative extrusion and Molten Metal Technology's CEP.

DuPont has entered into a working relationship with Molten Metal Technology and has funded development programs to use CEP to upgrade process streams of in-house waste at a number of its facilities worldwide.

For polymer-specific chemical recycling, the investment needed for the technology is certainly lower. But you have to count for increased costs on the collection and sorting aspect. The output is of relatively high value because these are monomers which can be directly repolymerized.

We have to include, in the overall economic overview, that the waste owner will contribute with diverted disposal costs. Once these considerations have been made for specific polymers, it may be that, to show a profit, we may need an additional subsidy from the government or the industry.

Now let's examine chemical recycling for mixed polymers. As we've seen, mixed polymers technologies need high investments overall. However, collection and sorting are both certainly easier and cheaper.

The value of the output is normally low, because we're talking about mixed oils or synthesis gases which compete with similar products coming from natural oil. Again, the waste owner has to contribute with diverted disposal costs and, in this case, the industry or government subsidy will be higher to make this technology economically attractive.

Overall, we can note for the future, an on-going improvement of sorting technologies, increasing oil prices and mounting disposal costs that will reduce and ultimately eliminate the need for external subsidies.

Energy recovery for rubbers is a valid option in cases where other recovery methods cannot be used for either ecological or economic reasons. Compliance with existing German and EC legislation confirms that modern fume treatment technologies used in modern energy recovery are absolutely safe, even for polymers or rubbers containing halogens.

Another energy recovery option is cement manufacture, where you need substantial energy which could be supplied via plastics and rubbers. Since the process operates at high temperatures from 1,600 to 1,800 °C, this ensures the gases or fumes are absolutely not harmful. There is also no solid waste left behind, since it stays in the cement.

I would now like to use this opportunity to bring you up-to-date on the activities of the organization I mentioned before, the Gesellschaft fuer Altgummi Verwertung System or "GAVS".

At the end of last January, GAVS completed a test with 50 tons of rubber waste - 50% tyres and 50% elastomeric parts - in the hydrogenation plant of Veba Oel Corporation in Bottrop. The 15-20% rubber was mixed with the heavy oil and then high pressure was applied at a temperature of 450 °C, with light oil and gas as an output. These outputs were used in refineries to make new products for the petrochemical industry. GAVS is currently negotiating to work with several hundred tons of feedstock.

In another pilot plant, a brown coal gasification plant in the "Neue Laender" was also successful. Adding methanol to the end product - synthetic gas - you can produce a series of new feedstocks for the petrochemical and chemical industries.

GAVS is currently working on developments with the ENRO corporation in Essen for recovery of energy for heating and electricity-generating plants.

I previously referred to the subject of cement manufacture. Two hundred thousand tons of rubber and plastic waste could be currently used in the cement production industry in Germany alone; GAVS is also working with the German Ministry for the Environment to get rubber recyclate accepted as a substitute for coal in this process for cement manufacture.

I would like to mention one final point to illustrate the active approach DuPont is taking to search for answers to the problem of rubber recycling.

DuPont, Veba Oel and RWE own a refinery in Karlsruehe. They have applied to the authorities for a permit to build a 1.2 million ton hydrogenation facility there, with the possibility of having plastic and rubbers mixed with heavy oil as a feedstock. This facility could start up as early as 1996, if permission to go ahead is granted.

In conclusion, rubbers, because of their cross-linked nature, cannot be treated like thermoplastics. Therefore, they can only be mechanically recycled to a limited degree and the recyclate used as fillers in less demanding rubber formulations.

However, rubbers could constitute an interesting feedstock for chemical recycling. Certainly, additional capacities of chemical recycling must be installed in the near future. The increasing disposal costs will also make this technology viable from an economic point of view.

In the meantime, the energy content of rubbers can be recovered in cement manufacturing facilities or in the production of heat and electricity in modern energy recovery operations.

As I stated in the beginning, the automotive industry – like others – is now under growing legislative pressure to provide concrete contributions to the issue of waste, plastic and other materials.

The PRAVDA scheme is, in my judgement, an approach which merits considerable attention from the industry, legislators and others, as the way forward for the dismantling, sorting and collection logistics – essential elements of any waste disposal process.

Polymers recycling will make an increasing contribution to the waste management process, particularly chemical recycling.

Chemical recycling technologies exist, are being improved and will play an increasingly important part in the total waste management equation.

And here, we – that is, the chemical industry – can do what it does best. For 70 to 80 years, the industry has devoted considerable effort to researching ways to combine monomers. It is perhaps ironic that we have finally learned to look at the problem the other way around: using all our accumulated expertise and experience to do the opposite.

But it is a way whereby we, too, can contribute to finding solutions to the waste issue.

It will take time to build up sufficient capacity in chemical recycling – partly because of the economics, but mainly because we have to assure a constant availability of quality feedstock.

Chemical recycling is not an instant miracle cure. But it is the preferred answer, long-term – a point we in the industry must make forcefully, both to legislators and to the public, to change their mistaken belief that mechanical recycling is always and automatically the best solution.

What is essential is that we judge each solution on a case-by-case basis and apply eco-balance analysis to ensure that we adopt the solution which is the most efficient and effective from an economic standpoint, and which has the lowest possible environmental impact.

In conclusion, I would like you to go away today remembering one thing. In Chinese, the word "crisis" is made up to 2 symbols: the first means danger, the second means opportunity.

That has a message for us all; the main challenge we face today is now to transform the public waste issue from a threat and a danger to a successful business opportunity. I have no doubt that we can and will.

New Technology for Waste Minimisation

Profit and Waste – Allied Allies

Malcolm J. Braithwaite

ALLIED COLLOIDS LTD, PO BOX 38, LOW MOOR, BRADFORD, WEST YORKSHIRE BD12 0JZ, UK

INTRODUCTION

This paper will show how profit can be generated from waste minimisation, not just through in-house efforts, but through developing products to help other companies to minimise waste. It will use as an example an organisation which has thrived on producing products for this purpose and will simply demonstrate the theme of this symposium and show how chemists can contribute to waste minimisation.

The National Economic Development Office Specialised Organics Sector Group set up a Task Force in 1991 to study environmental challenges and opportunities for the sector in the current regulatory and political climates. Waste minimisation and the development of products and processes to eliminate waste featured significantly in the final report titled "The Chemistry for a Better Environment" published in March 1992.[1] Aspects such as clean technology, IPC, biotechnology, government policy and strategies for good environmental practice will be covered at this symposium but this discussion will focus on the identification of environmentally driven product opportunities and illustrate from Allied Colloids' product portfolio how chemists can develop products such that profit can be made from waste minimisation. The contribution to a better environment will be clearly demonstrated.

ALLIED COLLOIDS EXPERIENCE

Allied Colloids has grown throughout the 1980s at a compound growth rate of 14% per annum largely generated by organic growth of existing businesses. This coincided with the worldwide awareness of the need to conserve energy and feedstocks post the oil crises of the 1970s. The three areas of most significance to this discussion are Pollution Control, Paper, and Mineral Processing Divisions. It is in these three areas where the driving factor of the customer industries was the minimisation of waste, both in the materials area and the energy field. Allied Colloids' performance chemicals are sold at not insignificant prices and the only real justification is the saving made in recovered or retained material and energy. Illustrating the point with a couple of specific examples later will demonstrate how the opportunities were identified and developed.

SIGNPOSTING TECHNIQUES

The identification of opportunities to profit from waste is not particularly easy but, in the current economic climate, it should be much easier to sell a product which potentially increases efficiency and hence reduces costs.

In a section of the NEDO report, Gerry Avis (formerly Albright and Wilson) and David Culpin (Chemical Industries Association) looked specifically at "Environmentally Driven Product Opportunities". They drew heavily from a report produced by the Aspinwall company for DTI entitled "Manufacturing Sector Environmental Study". [2] This survey studied the various environmental factors experienced by industries across the whole manufacturing sector and attempted to quantify their significance. The idea was to give chemical companies a 'route map' into chemical user industries to narrow the areas in which new product or process opportunities exist to solve environmental problems. For the purpose of this discussion the specific areas of waste quantity and waste hazard will be addressed.

Many of the problems of waste generation are inherent in the technology of the particular industry. Clearly a move to cleaner technology would be called for, but often the solution may be a processing aid, catalyst system, phase transfer agent, surfactant, etc. A major part of the challenge in spotting and developing the opportunity is in being aware of the customers problems in the first place. The 'signposting' technique can help to narrow the search and the derived lists may be a starting point. (See Figures 1 and 2).

INVESTIGATING OPPORTUNITIES

As indicated previously, the recognition that a problem exists for which a product solution may be developed is the first and most difficult step. An in-depth knowledge of the customer industry is necessary and clearly it helps if you already service that industry. However there are

FIGURE 1 – INDUSTRY SPECIFIC AREAS

INDUSTRY	WASTE PROBLEM	WASTE QUANTITY	WASTE HAZARD
Chemical	Many	H	H
Electrical/Electronic Eng.	Heavy Metals, chlor solvs	M	H
Leather	Heavy metals, sulphides	L	H
Mechanical Eng.	Soluble oils	H	H
Metal manufacture	Heavy metals, acids, oils	H	H
Motor vehicles and parts	Acids, oils, cyanide	M	H
Non-metallic minerals	Asbestos, cement, dust	H	L
Rubber/Plastics processing	Vulcanising accelerators	M	L
Timber and furniture	Phenols, formaldehyde	L	H

FIGURE 2 – PRODUCT SPECIFIC AREAS

PRODUCT AREA	WASTE PROBLEM	WASTE QUANTITY	WASTE HAZARD
Adhesives/sealants	Solvents, epoxies	M	M
Alkalis	Caustic, mercury, cyanide	M	M
Dyestuffs/pigments	Many	M	H
Paints/Inks/Varnishes	Solvents, colloidal carbon	M	H
Agrochemicals	Persistent residuals	M	H
Petroleum refineries	Sludges, catalyst residues	M	M
Tar/Bitumen	Sludges, phenols	M	M
Soap/detergents	Non-biodegradables, sulphur	L	M

ways to proceed if the objective is to create new product areas, not all of which are obvious. A few are listed for guidance:

1. Industry Associations - industry issues can be discussed in general terms and early identification of the important problems is possible.

2. Impending national and European legislation - nothing drives companies to seek solutions to problems faster than legislation. Monitoring industry - specific legislation usually identifies the key problem areas to be faced.

3. 'Green' organisations - public pressure can lead to companies reassessing their waste portfolios. Constructive discussion with 'green' activists can identify problems to be solved.

4. Developments in "leading edge" countries (US, Japan, Germany, Scandinavia) - a tremendous amount of information on waste minimisation products, processes and technologies is available in the open science-based press eg. Chemical Marketing Reporter, Japan Chemical Week, ECN etc.

5. Seminars and exhibitions - a rapidly expanding source of information and ideas.

6. Academe - interestingly it has been found that academics rarely recognise the true potential for their science. Regular and wide dialogue can be most productive.

7. Awareness of problems and solutions which could cross industry boundaries.

COMPANY STRENGTHS

Recognising the skills that a company can bring to waste minimisation is not always easy. However the most effective solutions are generated when company skills are well matched to the opportunities. Awareness of strengths can suggest solutions. Setting out strengths in a 3-dimensional array with the axes labelled: 1. Markets served   2. Processes operated   3. Materials handled defines the dimensions of competence quite well. Attempting to operate too far outside this matrix can be dangerous but should not really stop a firm from developing a new business or skill to deal with a novel situation. Ideally it is necessary however to match skills to likely solutions.

SPECIFIC EXAMPLES:

Having presented the theory, it is necessary to show how it works in practice.

The treatment of domestic effluent in municipal sewage treatment works is a major concern of Allied Colloids' Pollution Control Division. It is highly unlikely that the problem of ever increasing quantities of solid waste being produced by an ever increasing and better fed population will be dealt with via waste minimisation programmes but the chemist can alleviate the problem by facilitating more efficient disposal of the waste. Addition of cationic flocculants during the sludge dewatering process significantly reduces the volume of waste by allowing the removal of a greater proportion of the associated liquors. This gives considerable savings in landfill disposal costs and permits easier, more energy efficient incineration. Whilst this is not strictly a true example of waste minimisation it illustrates techniques by which waste can be minimised and allows demonstration of technologies crossing from one 'product' area to another. The obvious consequence of using this approach to waste systems containing suspended solid materials is that these materials can more easily be recovered and recycled thereby significantly contributing to the minimisation of waste.

This was particularly recognised by Allied Colloids' Mineral Processing Division in its dealings with the Coal Industry. Modern methods of mechanical coal mining using high output shearers have resulted in the amount of fine coal produced increasing from about 10% to about 25% of total run-of-mine production. These fines would be wasted to landfill unless they could be isolated in usable form. Mined coal is first washed to remove shale and the fines are also removed in this process. By addition of flotation reagents the fines can be separated from the shales. However, dewatering by vacuum filtration needs to be assisted by addition of flocculants, to assist sedimentation, and filter cake dewatering aids from the Allied Colloids range. Even then, some fine coal can still be very wet and unsuitable for blending with power station fuel. The solution came from another area of Allied Colloids' business, namely polyacrylic acid super-absorbent polymers for use mainly in disposable nappies. The polymer chemists developed a variation of the superabsorbent polymer which can be added to the fine wet coal in a high shear mixer. In the mixer the absorbent polymer progressively takes up the free moisture which until now caused the particles to stick together. By the time it leaves the mixer the fine coal is now free-flowing and ready to be blended with power station fuel.

Allied Colloids Paper Division also makes a considerable amount of its money from helping paper manufacturers to minimise waste during the paper making process. Allied Colloids polymer chemists developed a world-beating system, marketed under the trade name "Hydrocol", which received recognition with the presentation of a Queens Award for Technological Achievement in 1990. The "Hydrocol" system enhances the retention of pigments and fillers and enhances the water removal in the vacuum section of the paper-making machine whilst also ensuring enhanced retention of the cellulose fibres. Hence there are several consequent aspects to the minimisation of waste in paper production. Solids retention is increased by an average 15% with a consequent increase in speed of production by up to 20%. Steam used for drying the paper is reduced typically by 10% and solids content in the effluents reduced by 60%. The effluent COD is consequently reduced by around 20% giving savings in waste water treatment costs. Further benefits obtained by increasing retention include large reductions in the consumption of other wet end additives such as size (20% decrease) and starch (10% decrease). Clearly, Allied Colloids' polymer chemists have given the company a significant edge in this market through some clever chemistry.

SUMMARY

Identification of needs for waste minimisation techniques or technologies, both internally and in the external market place, is clearly a skill which can be applied to the benefit of any commercial enterprise. This paper has demonstrated how such needs can be recognised and how the solutions to the problems can be tackled by chemists to generate products which confer significant savings through minimisation of waste. Real examples, developed by polymer chemists at Allied Colloids, demonstrate that profits can indeed derive from waste minimisation.

REFERENCES

1.   "The Chemistry for a Better Environment" - National Economic Development Office, March 1992

2.   DTI "Manufacturing Sector Environmental Study" - Aspinwall & Company, HMSO, 1989

The Role of Biotechnology in the Development of Clean Manufacturing

Stephen C. Taylor

ENVIRONET, ZENECA BIO PRODUCTS, PO BOX 2, BELASIS
AVENUE, BILLINGHAM, CLEVELAND TS23 IYN, UK

1   INTRODUCTION

It is perhaps not surprising that the growing demand for
processes and products that are more environmentally
benign than their predecessors, is leading to increased
interest in the potential for biotechnology to address the
need.   The ability of biological systems to perform
complex reactions without resort to extremes of pressure,
temperature or chemistry is well known. When combined
with an almost unique capability to carry out such
reactions with a very high degree of specificity and
selectivity, it provides the basis for a valuable
manufacturing technology and source of novel products
and effects.

The impact of new biotechnology is at several different
levels. These can be considered in three groups:

-       enhanced end of pipe treatments
-       novel technology for existing products and
        processes
-       product innovation

This paper will examine examples from each of these
groups moving 'up the waste pipe'.

2   END OF PIPE TREATMENT

Although biotreatment is a well established process for
effluent treatment this has generally been considered to be
a standard technology with little flexibility for innovative

and imaginative incorporation into total process flowsheets. However, effluent biotreatment is essentially concerned with the bioconversion of chemicals and is thus amenable to many of the approaches now being successfully used for the production of chemicals by biotechnology. The key difference lies in the desired products being carbon dioxide and water rather than higher value chemicals!

The impact that modern, intensive biotreatment can have on processes is well demonstrated by the use of Deep Shaft technology as an integral part of the new Pure Terephthalic Acid manufacturing plant in Taiwan. In this situation treated effluent is recycled to the plant in a cleaner form than the incoming water.

One of the growing concerns with end-of-pipe treatments is the level of organic, but relatively recalcitrant material that can pass through a plant untreated. Approaches to this problem such as charcoal adsorption or chemical oxidation have been demonstrated but such systems have a significant cost associated with them and often poor overall efficiency. An alternative approach is to recognise that most of these compounds are still biodegradable and the core of the problem is that the effluent biotreatment plant is not operating in a way that allows expression of the latent ability of certain micro-organisms to deal with these less degradable chemicals.

This approach has been used by our network to address issues including dichlorobenzene in chemical effluent, colour from dye manufacturing sites and halogenated aromatics from wood pulp bleaching plants. Some degree of specific customisation has been required in these cases but the validity, efficacy and cost effectiveness of the approach confirmed. It is important to recognise, however, that this is not simply a case of selecting the appropriate micro-organisms. Combination with effective, robust but often innovative process technology is the key. This is a good demonstration of how cleaner manufacturing processes can be achieved by seeing the effluent treatment system as part of the total process and not a 'bolt-on'.

3   NOVEL PROCESS TECHNOLOGY

Biotechnology can be very effectively used to derive cleaner methods for making existing industrial products or to provide ways of upgrading current products.

The manufacture in Japan of acrylamide is a good example of the former. In this case the conventional process utilises a copper based catalytic method. Although very effective the formation of by-products and residual copper are issues of growing environmental importance. An enzyme process has been developed and operated by the Nitto company based on the hydration of acrylonitrile. This quite remarkable enzyme provides access to high purity acrylamide at low temperature, high productivity and with no by-product formation: a good illustration of how biotechnology can replace metal based catalysis.

As an alternative to deriving totally new synthetic strategies as above, biotechnology can also be used to modify existing processes to overcome specific issues. The 'Silver Bullet' concept as it has become known, has for example been used to remove residual acrylamide from polyacrylamide by Zeneca. Here small amounts of monomer remain after polymerisation which raises concern because of the known toxicological effects of acrylamide. An amidase enzyme has been developed which converts acrylamide to acrylate without any side reactions on the polymer. The robust enzyme system works over a wide range of temperature and pH and can thus be incorporated directly into the polymer manufacturing process to derive polyacrylamide with non-detectable levels of acrylamide.

Another example of this concept concerns the removal of odour from paint, much of which arises from residual esters. An enzyme system developed by Zeneca can effectively hydrolyse esters directly in a paint or resin environment and can thus again be incorporated directly into existing plant operations to give low odour paint products.

Through this type of approach benefits can be realised from biotechnology without the need for substantial investment in new capital.

4   PRODUCT INNOVATION

The discussion to date has focused upon how biotechnology can be used to improve the overall performance of existing processes or derive 'cleaner' ways of producing existing products. The long term goal, however, has to be to utilise the opportunity presented by bio technology to approach problems from new angles and develop new product concepts.

The pulp and paper industry has already been mentioned as a source of halogenated aromatics arising in pulp bleaching effluent due to the use of chlorine based compounds in bleaching. An approach to this problem has been to utilise specific micro-organisms, selected for their ability to degrade halogenated compounds and incorporate them into bio-treatment plants. The better longer term approach, however, is to move away from the source of the problem and seek an alternative to chlorine for pulp bleaching. In this process a precise reaction is required to liberate lignin from the cellulose, a situation where the specificity of enzyme technology is of great value and offers significant advantage over the non-chlorine, less specific, alternatives such as ozone. This requirement has lead to our development of the ECOZYME range of enzyme based pulp bleaching catalysts. These allow the industry to move away from the extensive use of chlorine chemistry without incurring the substantial capital investment that alternatives require.

This represents a good example of how the use of precise biological catalysis, customised to meet the needs of a particular problem can allow a move away from the stoichiometric use of chemicals, a major factor in the advance of clean manufacturing. This is of particular value where biological materials are involved in processing as in pulp and paper. However, this is not a constraint and this paper has hopefully illustrated the wider potential of biotechnology to aid the development of environmentally acceptable processes and offer new product opportunities.

The Benefits of Computer Control in the Process Development Laboratory

R. H. Valentine

HIGH FORCE RESEARCH LTD, UNIT 1D, MOUNTJOY RESEARCH CENTRE,
STOCKTON ROAD, DURHAM DH1 3SW, UK

Introduction.

Plant scale computer control in the fine chemicals manufacturing industry is becoming increasingly common. However its use in research and process development laboratories is not so common and tends to be used primarily for safety and hazard assessment.

We have found at High Force Research that, used imaginatively, such equipment can be of benefit to process R&D chemists, in many ways, to help study and optimise chemical reactions both quickly and efficiently. A number of such examples will be given which will illustrate ways in which reactions can be optimised and hence minimise materials, effluent, plant and energy usage.

Equipment.

A small number of reaction calorimeters are commercially available which are suitable for this type of investigation but they are expensive and require trained personnel to use them to best advantage. Simpler systems can be built around standard Personnel Computers with suitable off the shelf electronic components and laboratory equipment but this does require a substantial degree of effort by in house electronic engineers, programmers and chemists to build a reliable and workable system.

High Force Research has developed a system which is both flexible in terms of laboratory equipment that can be connected to it and is very easy to use, requiring the minimum of training. It is commercially available and is affordable costing about the same price as a good quality HPLC system.

The minimum requirements for a usable system are outlined below.

i)   2 temperature inputs.
ii) Industry standard general input device e.g. 4-20 milliamp interface, to receive other measurements such as those available from pH meters, Oxygen meters, pressure transducers etc.
iii) Electronic interface to a suitable heater/cooler circulator for use with a bottom outlet jacketed flask.
iv)   Interface to balances.
v)   Interface to pumps or other reagent addition devices such as nitrogen displacement solenoid valve assemblies which are useful for controlled addition of air sensitive or other problem reagents that are difficult to pump.

The ability to datalog all reaction parameters, as well as controlling reactions, is a very important function as this enables post experimental analysis to be performed and a great deal of important information can be obtained with the minimum of effort. A graphical output of the reaction parameters has been found to be the best way of viewing the data. Graphs are best displayed in colour although the examples shown in this article are of necessity displayed in black and white. Reagent weights and graph scales have also been omitted for clarity.

Examples of use.

a) Temperature control.

This is perhaps the most fundamental experimental parameter that needs to be controlled and logged. Most reactions are temperature sensitive and often manual control is far from satisfactory. A great deal of information can be gained about reactions just by logging the extent of any exotherms or endotherms associated with a particular reaction.

The first example (fig i) shows an exothermic halogenation of an olefin in which the halogenating reagent and olefin are simultaneously pumped into the reaction vessel containing an inert solvent. The extent of the exotherm is shown by the differential between the internal pot temperature (upper solid line) and the jacket temperature (lower dashed line). The two diagonal lines indicate the addition of reagent and olefin. As soon as the addition is complete the exotherm stops and the jacket temperature is automatically increased to bring the pot temperature back to the reaction set point.

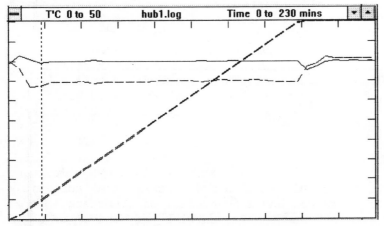

| T'C 0 to 50 | hub1.log | Time 0 to 230 mins |

Fig(i).   Halogenation of an olefin.
This example illustrates that under the conditions used reaction is immediate and stops as soon as the reagents have been mixed.

The second example (fig ii) shows a reaction in which the exotherm is milder and continues after the reagent addition is complete. The point at which the exotherm completely dies away indicates the time required for the reaction to take place.
The measurement of an exotherm can also be valuable in indicating whether the correct quantity of reagent has been used. An experiment (fig iii) which had previously been done conventionally with ice/water cooling and manual addition of reagent, clearly shows that the exotherm associated with the reaction dies away some time before the addition of reagent is

complete, indicating that the excess of reagent being
used was probably not necessary.

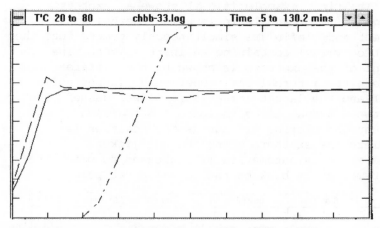

Fig(ii). Extended Exothermic Reaction.

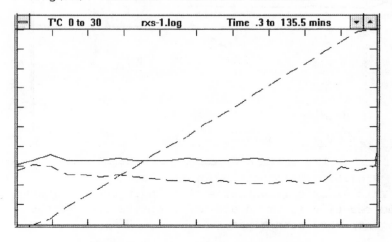

Fig(iii). Use of excess reagent.

The use of temperature ramping control can be
extremely useful in crystallisation work. Recently we
were required to produce a high purity, crystalline
material which had a melting point of 25'C. When we
crystallised this material from a suitable solvent
using an ice/water bath it was found that the
filtration was very slow, resulting in the solid
melting in the funnel and producing a low yield of
impure material. The slow filtration was due to small
crystal size which could not be controlled by cooling

in a conventional ice/water bath. When we cooled the
solution using a controlled temperature ramp, we
found that the material crystallised out at 5'C, as
indicated by the exotherm shown in fig (iv) which was
due to latent heat of crystallisation. When the
material was cooled to and maintained at 5'C (fig v),
it was found that the crystallisation was much better
resulting in a larger crystal which filtered rapidly
giving a good yield of pure material.

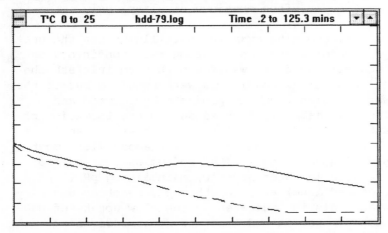

Fig(iv). Cooling with controlled temperature ramp.

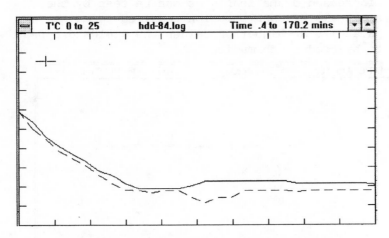

Fig(v). Controlled crystallisation at 5'C.

The same technique has been used successfully
with other crystallisations.

b) pH Control.

The effect that pH can have on reactions can be
quite significant. A great many chemical reactions
are subject to pH changes and maintaining an optimum
pH condition can be extremely difficult with the
normal manual methods available. However if computer
control is employed then optimum pH conditions can
easily be maintained and in addition valuable
information is often obtained simultaneously.

An interesting example that illustrates the use
of pH control was a condensation reaction involving
chloroacetic acid. It was found that to initiate the
reaction alkaline conditions were required but if the
pH was too high then the products hydrolysed and
yields were markedly lowered due to the formation of
unwanted by-products. Figure (vi) shows a reaction in
which the condensation reagent was added (left most
dashed diagonal line) to the sodium salt of chloroacetic acid in water at 65'C, maintaining pH (top
most dotted line) at 9. At this pH, reaction was very
slow and this is indicated by the slow uptake of NaOH
solution (lower dashed line) to neutralise the
chloride ion released. When the pH was adjusted to 11
at the end of the reagent addition then the reaction
rate increased dramatically as can be seen by the
faster uptake of NaOH solution required to maintain
the pH at 11. At higher pH values the yields dropped
due to by-product formation.

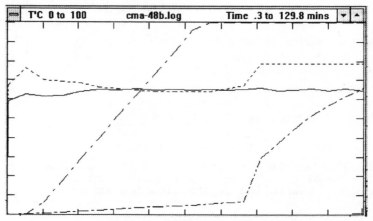

Fig(vi). Condensation reaction with Chloroacetic acid

The reaction profile shown in Figure(vii) illustrates the same reaction done at pH 10.5 and 70'C.

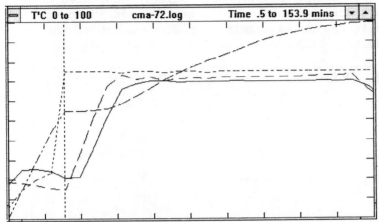

Figure(vii). Chloroacetic acid condensation reaction at pH 10.5 and 70'C.

This reaction profile shows the initial formation of the sodium salt of chloroacetic acid by the addition of sodium hydroxide to bring the pH to 10.5. The solid line is the internal pot temperature, the dashed line that follows it is the reactor jacket temperature; pH is indicated by the upper dotted line and the addition of sodium hydroxide is shown by the curved dashed-dotted line. An initial strong exotherm is shown as the chloroacetic acid is neutralised and the pot temperature rises from 20'C to 25'C. The automatic addition of NaOH solution stops once the pH reaches the set point of 10.5. At this time the other reagent is added in one portion which is indicated by the vertical dotted line marker and then the temperature is increased to 70'C. Once the temperature passes 50'C the condensation reaction starts and in order to maintain pH at 10.5 the NaOH solution is automatically pumped in.

An indication of the rate of reaction is shown by the NaOH uptake. The reaction is assumed to be complete when no more NaOH is required and then the reaction mix is cooled down.

The speed of uptake of the pH control reagent can give a good indication of reaction rate at different temperatures. Figures (viii) and (ix) are

the reaction profiles of two experiments in which an
aromatic amine is methylated at 5'C and 20'C with
dimethyl sulphate. Sulphuric acid is generated as a
by-product and unless it is neutralised in some way,
then it can cause significant degradation of the
product.

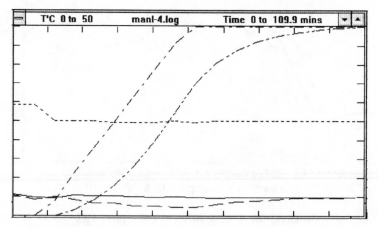

Fig(viii). DMS methylation of amine at 5'C

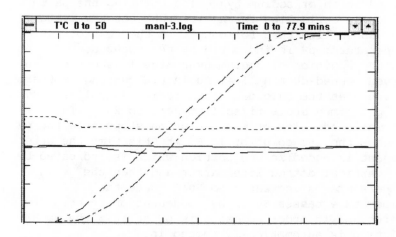

Fig(ix). DMS methylation of amine at 20'C.

The dimethyl sulphate (DMS, leftmost dashed
line) was added over 45 minutes in each case,
maintaining the pH at 7 by automatic addition of base
(rightmost dashed line). It can clearly be seen that
reaction at 5'C was considerably slower than at 20'C
as the base addition, which is a direct indication of
reaction rate, takes longer than in the latter case.

Reaction at 20'C is probably near optimum as the base addition finishes only a few minutes after the end of the DMS addition. The exotherm is also extended from 45 minutes to 70 minutes in the 5'C reaction.

pH control can be useful in many other ways. By selective adjustment of pH, precipitation of a desired material can often be achieved. It can also help in the separation of isomers which have different pKa values and the control and display advantages of a computer system make this type of separation very much easier.

c) Comparison of catalysts and optimisation of catalyst loadings.

As all experiments done using a computer controlled reactor can be accurately repeated then it is an ideal way in which to compare the effect of different catalyst grades, types or loadings as all other experimental parameters can be kept constant while the catalyst type or loading is varied.

The example illustrated in figures (x and xi) show two identical reactions catalysed with 1% and 3% respectively of catalyst. In both cases an acidic by-product was produced and a measure of the reaction rate is illustrated by the take-up of a basic pH control reagent.

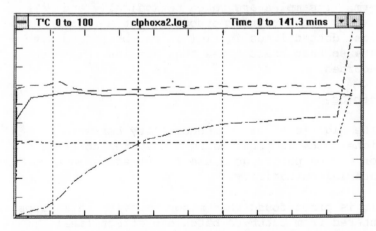

Fig(x). Reaction using 1% by weight of catalyst.

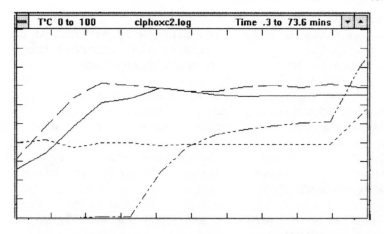

Fig(xi). Reaction using 3% by weight of catalyst.

One of the reagents is added in a single portion to a mixture of catalyst, solvent and other reagent. It can clearly be seen that the uptake of pH control reagent (lower curved dashed line) is much faster in figure (xi),in which a larger loading of catalyst was used, and also the exotherm is more pronounced (the jacket temperature indicated by the upper dashed line drops further towards the internal pot temperature).

The same techniques can be used even if there is no direct measure of reaction rate, except by exotherm, if samples are taken periodically and analysed by conventional methods. Event markers (vertical dotted lines in fig(x)) can be logged with the data so that exact times that samples are taken are recorded.

d) Other Uses.

The high level of reproducibility between reactions which is found using a computer controlled reactor can be put to good use in the assessment of raw material suitability.

It is often found that minor impurity variations encountered from batch to batch can effect final product quality. This is often the case with products with a high colour specification and small impurities in the raw materials can degrade quality and lead to expensive and wasteful reworking.

In addition to normal analytical testing an end use test has to be used in which the final product is synthesised from the new feedstock and then its colour quality assessed. However using conventional synthetic techniques there is often no guarantee that every experiment is repeated in exactly the same way and so there is often an element of doubt as to whether quality variation is due to the raw material quality or variations in laboratory manipulation. Using a computer controlled reactor, reaction profiles from different experiments can be compared giving much greater confidence in the assessment.

Conclusion.

This brief review of the benefits of computer control in the process development laboratory has tried to illustrate some of the advantages gained from the markedly superior control and recording of experimental parameters in laboratory synthesis.

It can be extremely useful as an investigative tool during process optimisation often showing reaction detail that is not apparent with conventional methods.

It is ideal for doing step by step comparison work with different raw materials or catalysts and gives far greater confidence in results and generates a permanent record of experiments.

An added advantage not mentioned so far is the productivity gains of laboratory staff. This can be particularly high with repetitive, "handle turning" operations and can free valuable, trained personnel for more useful and interesting work. It has been our experience that with a simple to use, flexible system, worker satisfaction is often increased.

The examples illustrated have concentrated on temperature and pH control but other parameters can be controlled if the correct instrument is interfaced i.e. ion selective electrodes, oxygen meters, pressure transducers and colorimeters are all

suitable and are relatively inexpensive. The data
gathered is of a very detailed form and makes it
amenable to analysis by computer batch simulation
programs often used by chemical engineers.

The system developed at High Force Research has
been used to great benefit in process optimisation
and has helped to maximise yield, plant occupancy and
product quality as well as minimising by products and
effluent in a number of manufacturing processes.

The Use of Heterogeneous Catalysis in the Synthesis of Fine Chemicals and Chemical Intermediates

James H. Clark

DEPARTMENT OF CHEMISTRY, UNIVERSITY OF YORK, HESLINGTON,
YORK YOI 5DD, UK

1   INTRODUCTION

The worldwide market for catalysts is $ 5000m per annum with the
product value dependent on them being $ 240,000m per annum. It
has been estimated that 90% of the important new processes
introduced since 1930 depend on catalysis. Rather than being close
to saturation however, there is significant market growth potential
due to (i) strong growth for environmental (e.g., automobile)
catalysts and (ii) the need to improve existing catalytic processes
(e.g. to improve reaction selectivity and to replace toxic catalysts)
and to apply catalysis in areas where it has not traditionally played
a role. The crucial factor in the future growth of catalysis will
undoubtedly be "enviro-economics" driving industry towards new
more environmentally friendly products and processes.

One important aspect of "Clean Technology" will be the
increasing use of "Environmentally Friendly Catalysis" typically
involving the use of solid catalysts which lead to minimal pollution
and waste and which themselves are environmentally benign. The
application of such catalysis to fine chemicals and chemical intermediate manufacturing - an area where heterogeneous catalysis has
traditionally been little used - is likely to be especially important in
the future. The design and use of inorganic supported reagent and
molecular sieve catalysts in liquid phase organic chemistry will be
described so as to illustrate this rapidly developing field.

2   MICROPOROUS AND MESOPOROUS INORGANIC
CATALYTIC MATERIALS

High surface area, porous inorganic materials including alumina,
silica gel and clays are well established in many areas of chemistry
including chromatography, and catalysis. The catalytic potential of
such materials is based on a number of properties including:
(i)   high surface area providing a large number of active sites

(ii)   complex surface chemistry including Lewis acid and base
         sites and Brönsted acidity and, in some cases, the presence of
         exchangeable cations.
(iii)   physical irregularities including pores, cracks, and edges
         where reactivity and sometimes selectivity can be enhanced.

The subject of inorganic materials catalysis has been greatly
enhanced by two significant developments: the synthesis and
application of an increasingly wide range of so-called molecular
sieves with molecular-dimension pores capable of shape selective
chemistry and the design and application of supported reagents
where new active sites (ions or molecules) are dispersed over the
surface or in the pores of a porous inorganic material.

Zeolitic materials are able to recognise, discriminate and
organise molecules with precisions that can be less than 1Å. This
molecular sieving ability along with their regular structure and
inherent acidity has made them popular catalysts for many large-
scale, shape-selective hydrocarbon reactions. Typically, zeolite
catalysis has involved acid-catalysed reactions or bifunctional (acid-
metal) processes on quite small molecules. New developments in
molecular sieve design open the door to a broader range of
reactions as a result of new extra-large pore materials (pore sizes of
ca. 14Å - almost double that of the largest aluminosilicate zeolite -
are now available) and materials designed for other than acid
catalysis - including base catalysis, asymmetric catalysis and
oxidation catalysis.

Shape selective catalysis using a zeolite was first described in
1960 and the first inorganic supported reagent is attributed to an
article published in 1968. Although supported reagents can be
based on microporous zeolites, they are more commonly based on
mesoporous supports such as silica gels and aluminas. These more
open structures are often more amenable to catalysis in liquid phase
organic reactions. High local concentrations of reagents in the
internal "micro-reaction chambers" can lead to synergistic support-
reagent effects that lead to very high activities. The larger pores are
however, less likely to promote shape-selective catalysis although
adsorption mechanisms and partitioning effects between the highly
polar internal regions of the material and the typically weakly polar
liquid phase can lead to product selectivity and unusual trends in
substrate activity.

Clearly a compromise between the high degree of shape
selectivity in zeolitic materials and the often better activity of larger
pore materials could be valuable. In this context, the new larger
pore molecular sieves including pillared clay materials offer much
promise for future development.

3   ENVIRONMENTAL FACTORS

The potential advantages of using molecular sieves and the related support reagents in organic synthesis are numerous and include:
*   improvements in the rate and/or selectivity of a reaction compared to conventional conditions
*   ease of handling and use of generally non-toxic materials (active sites are generally located internally)
*   good thermal and mechanical stabilities (compared to organic/polymer based materials)
*   reactions are usually carried out in non-polar solvents or, where possible, in the absence of solvent
*   ease of product and reagent recovery (the materials can be filtered from the reaction mixture and can often be reused).

These advantages are especially important in today's "green" technological climate where waste resulting from inefficient reactions and the use of toxic reagents are becoming increasingly unacceptable. In the light of this, there is increasing interest in the application of molecular sieve-type catalysis to the manufacture of fine chemicals and chemical intermediates especially where improved reaction efficiency (e.g., through more selective chemistry) or the replacement of unacceptable reagents are desirable.

4   CASE STUDIES

Friedel-Craft reactions are excellent targets for new, cleaner chemistry since traditional methods:
(i)   use aggressive reagents such as $AlCl_3$, BF_3 and HF
(ii)   often lead to high molecular weight by-products
(iii)   can produce relatively large quantities of corrosive and sometimes toxic effluent.

The important of the reactions which include alkylations, acylations and sulphonylations, in areas such as pharmaceutical intermediates, cosmetic and paper coating products adds to the importance of seeking new chemical methods.

In the manufacture of benzophenones for example, the most commonly used catalyst is $AlCl_3$, yet the strongly complexing nature of this Lewis acid results in greater than stoichiometric quantities being required. A simplified mechanism can be written as:

$$ArCOCl + AlCl_3 \longrightarrow ArCO^+ [AlCl_4]^-$$
$$Ar'H + ArCO^+ [AlCl_4]^- \longrightarrow [ArCOAr'.AlCl_3] + HCl$$
$$[ArCOAr'. AlCl_3] + xH_2O \longrightarrow ArCOAr' + AlCl_3. xH_2O$$

The final aqueous work-up step (to destroy the product- $AlCl_3$ Lewis complex) leads to acidic aqueous aluminium-containing effluent which will become increasingly difficult to dispose of especially in view of the possible link between aluminium and Alzheimer's disease.

Great progress has been made in recent years in finding more environmentally friendly alternatives to $AlCl_3$ and molecular sieve-type catalysts lead the way.

Ion-exchanged montmorillonites and zeolites have been used to catalyse Friedel-Craft acylation and alkylation reactions but the activity of such materials is usually too low so that catalytic acylations using carboxylic acids for example, are restricted to the more reactive longer-chain acids:

$$ArH + CH_3(CH_2)_n CO_2H \xrightarrow[-Al^{3+},Ni^{2+}(etc)]{clay, zeolite} ArCO(CH_2)_n CH_3$$

The real breakthrough has come with the development of some montmorillonite based supported reagents such as "clayzic" which have been successfully applied to Friedel-Craft alkylations, sulphonylations, benzoylations, and to a lesser extent, acylations. The effectiveness of the supported reagent catalysis can be related to the strength of the electrophile (R^+, RSO_2^+, $ArCO^+$ or RCO^+) and to the nature of the acid catalysis required enabling the catalyst to be designed for the particular applications. Alkylations catalysed by clayzic are an excellent example of the synergistic effect of a molecular-sieve type support and a normally weakly active reagent:

$$ArH + Ar'CH_2Cl \xrightarrow{catalyst} ArCH_2Ar' + HCl$$
$$\text{catalyst = clayzic} \gg \text{montmorillonite} > Zn^{2+}$$

It is now believed that this synergism is due to the high local concentration of Zn^{2+} ions in silica mesopores created from the clay structure giving an unusual example of an essentially pure solid Lewis acid catalyst. It is likely that these mesopores are cylindrical in shape and therefore unlikely to impose any significant geometrical constraints on substrate or product molecules. Unexpected trends in substrate activity are, however, observed. Thus halobenzenes are surprisingly more reactive than benzene towards hydrated clayzic catalysed benzylation - an effect opposite to that observed in homogeneous (e.g., $AlCl_3$) catalysis. These and other results help us to build up a picture of these remarkable catalytic materials, based on internal reaction vessels with diameters of ca 100Å and very high local reaction fields which can repel low polarity/polarisibility molecules. Remarkably, relative rates of substrate alkylation may be controlled by the hydration level of the catalyst - this may open the door to designing catalysts which can sieve substrate molecules on the basis of electronic rather than geometric factors.

Oxidation chemistry is another area where the need for new catalytic methods has been enhanced by environmental considerations. Selective oxidations of alkylbenzenes for example, to give useful acid, aldehyde and ketone products are normally difficult reactions and can require environmentally unacceptable reagents such as Cr(VI), Mn(VII) or corrosive cobalt-acetic acid-bromide systems. Some progress in the design of molecular sieve-type catalytic materials has been made. Alumina supported chromium - a complex material based on a small number of highly efficient chemisorbed chromium centres - will enable selective oxidations to be carried out with excellent turnover efficiency and at low cost (air as the consumable source of oxygen):

$$ArCH_2Ar' \longrightarrow ArCOAr'$$
$$ArCH_2CH_3 \longrightarrow ArCOCH_3$$

Rates of reaction are rather low, however, and more active solid catalysts that maintain selectivity (ie. avoiding over-oxidation) are needed for some applications.

Other valuable molecular sieve-type oxidation catalysts include a titanium containing zeolite-based material, TS-1, which is used commercially to produce catechol and hydroquinone from phenol and aqueous H_2O_2. TS-1 is actually a broad-range catalyst for oxidation reactions. Vanadium substituted into alumino-phosphate and zeolite materials can give catalysts suitable for the liquid phase oxidations of alkylaromatics and alkanes. Pillared clay materials also show promise in this area.

Among the many challenging areas are asymmetric synthesis, base catalysis useful for many organic reactions (including C-C bond-forming reactions), halogenation and nitration of aromatic rings (especially where more site-selective monofunctionalisation can be achieved) and aromatic substitution reactions. The potential for using molecular sieve-type catalysis in liquid phase organic reactions such as these is enormous and there are now great opportunities for developing new, more environmentally acceptable processes based on such materials.

GENERAL REFERENCES:

1   Supported Reagents: Preparation, Analysis and Applications, J.H. Clark, A.P. Kybett and D.J. Macquarrie, VCH, New York, 1992.
2   Solid Supports and Catalysts in Organic Synthesis, K. Smith, ed., Ellis Horwood, 1992.
3   New Reagents for Old Processes, T.W. Bastock and J.H. Clark, Chapter in Speciality Chemicals, B. Pearson, ed., Elsevier, London, 1991.
4   New Vistas in Zeolite and Molecular Sieve Catalysis, M.E. Davis, Acc.Chem. Res., 1993, 26, 111.

The Creation of New Catalysts

C. J. Suckling

DEPARTMENT OF PURE AND APPLIED CHEMISTRY, UNIVERSITY OF STRATHCLYDE, 295 CATHEDRAL STREET, GLASGOW G1 1XL, UK

1 CATALYSTS AND CLEAN SYNTHESIS

I was once challenged by Professor Jack Baldwin that to create a new catalyst (an enzyme in the context of the discussion at the time) was a much greater challenge to the synthetic chemist than to inhibit one, as a medicinal chemist might wish to do. This paper responds to the challenge by discussing the place of catalysis in what has been called *clean synthesis*[1] and illustrating some aspects of the field with reference to our work. The term *clean synthesis* was used in the context of a report to the AFRC/SERC Clean Technology Unit to characterise chemical reactions that led to compounds that were commercially valuable through the effects of their applications (effects chemicals) with minimal energy consumption and waste. For this purpose of the Report and this paper, both biological and non-biological catalysts can be considered.

There has been a conventional wisdom in some quarters that biological catalysts have intrinsic advantages general over conventional chemical catalysts. However, as was argued[1], the choice of the best catalyst for the job depends upon an evaluation of the details of the problem in hand which must clearly include separation and isolation operations. On the other hand, there are a number of general criteria that give a guide to the properties required of a system that qualifies for the title of clean synthesis. To quote just two from the report:

> *High yields must be obtained based upon all reactants, not just upon the most expensive ones. Unless a simple excess can be recycled, all other reactants will become at best by-products and at worst waste.*
> *Reagents not incorporated into the desired products should be avoided.*

The practical expression of these obvious guidelines might, for example, be developed through the discovery of new catalysts and by the avoidance of conventional activating and leaving groups in organic synthesis, in particular reducing the use of halogens in industrial organic chemistry. At Strathclyde, we have been concerned with two approaches to devising new chemistry that approaches the stringent requirements of clean synthesis namely *biomimetic chemistry* and *new biological catalysts*.

2 BIOMIMETIC CATALYSTS

Breslow[2] was the first to draw detailed attention to biomimetic catalysts, that is catalysts that appropriate some of the features of enzymic catalysis, such as selective binding, and apply them to an non-biological reacting system. Our attention had focussed upon aromatic substitution, in particular on ways of avoiding excessive reactivity or toxicity in reagents. From studies of micellar systems, we had found that selective chlorination of phenols was possible and nmr spectroscopy had indicated a preferred average orientation of phenols in a micelle[3]. These observations led to investigations of pyridinium salts and aromatic nitration[4]. Nitration was chosen because of the potential of converting the nitro group into other functional groups and the wish to minimise the use of strongly oxidising and acidic electrophiles (Figure 1).

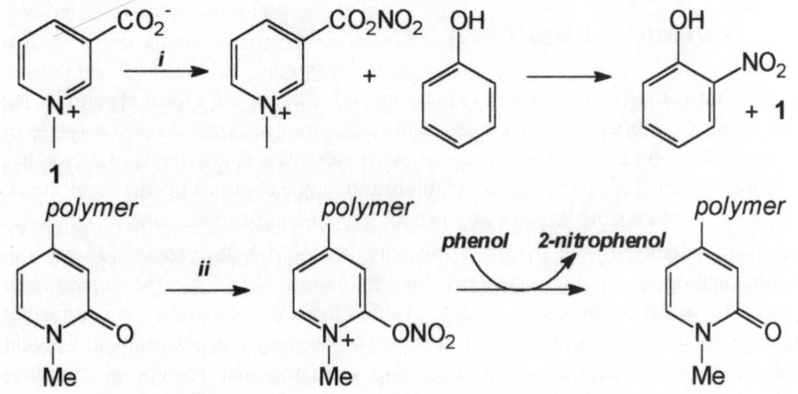

Figure 1. Biomimetic selective preparation of 2-nitrophenol.

The source of nitrogen in the most developed system was NO_2/N_2O_4 which reacted with the carboxylate of the pyridinium salt to form essentially a mixed anhydride; the reagent was thus trapped and its reactivity modified according to the properties of the carrying group, the pyridinium salt. In solution, this reaction afforded good yields of 2-nitrophenol selectively. We were able to deduce from spectroscopic studies (ir, nmr) that the selectivity was due to a donor-acceptor interaction between the pyridinium and phenol rings together with a specific hydrogen bond between the phenol and the nitrating functional group in such a way that 2-substitution was strongly preferred.

In principle, the pyridinium carboxylates **1** could be recovered from solution and recycled but this was cumbersome. To avoid extraction problems, the natural step of immobilising the system upon a polymeric support was taken. The best approach to this was found to be through derivatives of 2-pyridone, probably the smallest biomimetic molecule yet examined. We found that a suspension of the polymer derived from 4-vinylpyridine could be activated with NO_2; the introduction of phenol into the solution led to essentially quantitative 2-nitration, the product being isolated in crystalline form simply by filtration and evaporation of the solvent. The polymer could then be recycled.

Of course this is not a catalytic reaction, it is merely a stoichiometric reaction with a recyclable, solid-supported reagent. However it does open up a new approach to selective substitution of reactive aromatic compounds in which polysubstitution is to be avoided; the only by-product observed in this reaction was less than 1% 4-nitrophenol. Selectivity was clearly of importance in this study but, if the nucleophile is changed from phenol to an alkene or alkyne, there are also possibilities for extending this approach into reactions that would be the equivalent of organometallic addition reactions leading to extended carbon skeletons. Such chemistry has the potential to avoid the introduction and subsequent displacement of halogen atoms by providing an alternative method for the activation of carbon chains.

3 CHEMICALLY MODIFIED ENZYMES

Every chemist using biocatalysis wants to have an enzyme with ideal specificity for the reaction of interest but, of course, nature has evolved only a limited range of enzymes. To obtain new enzymes or enzymes with modified functions has therefore become a major goal of bioorganic chemistry. On the one hand, modifications of existing enzymes by genetic engineering can be considered as has been clearly demonstrated by Holbrook with lactate dehydrogenase[5]. Even with this outstanding work, the enzyme retained its natural chemistry, the reduction of 2-oxoacids, whilst the acceptable range of side chains was expanded. It is a further challenge to obtain an enzyme that catalyses a selected reaction that is different from the naturally occurring reaction; also reactions that require electrophilic catalysis cannot be catalysed by proteins without the aid of cofactors. A few years ago, Kaiser investigated a number of modifications of papain[6] taking advantage of the reactive thiol group at the active site to introduce cofactors such as flavins to catalyse oxidation reactions. With the need for electrophilic catalysis in the formation of carbon-carbon bonds in mind, we have followed this work up to gain experience in the possibility of obtaining preparatively useful catalysts[7] (Figure 2).

Analogues of nicotinamide dependent enzymes were prepared but, although they mediated the stereoselective reduction of electrophilic substrates, the reaction rates were very low and ^{13}C nmr studies indicated that a long-lived stable intermediate was being formed. More effective were analogues of thiamine dependent enzymes. The alkylation of papain with bromomethyl thiazolium salts after extensive purification afforded proteins that catalysed transformations of 2-oxoheptanal; this substrate was selected with the expectation that cyclisation would occur. It was therefore surprising to find that the major product was the acetoin-type dimer of 2-

oxoheptanal and that a smaller proportion of 2-methylcyclohept-2-enone was produced[8]. Although detailed kinetics have not been carried out, an estimate of the rate enhancement of the thiazolopapain is 2000-fold with respect to the free thiazolium salt under identical conditions. Given a receptive environment, therefore, analogues of naturally occurring cofactors can be introduced productively into enzymes; such incorporation might well be important in the development of catalytic antibodies for synthetic reactions.

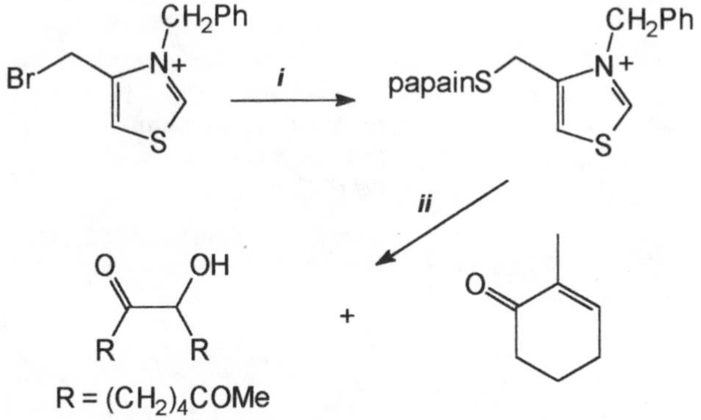

Figure 2. Preparation of and reactions catalysed by thiazolopapain. Reagents: *i* papain, pH 7.0; *ii* 2-oxoheptanal, pH 7.5

3 CATALYTIC ANTIBODIES

The scientific basis of the field of catalytic antibodies is well established. Although most studies have concentrated upon mechanisms of catalysis by proteins and upon hydrolysis reactions, much of our work has concerned the approaches to the creation of catalysts with potential in preparative chemistry. We have information so far on two series of antibodies, one of which was designed to catalyse aromatic substitution[9], and the other Diels Alder cycloaddition[10]. Neither reaction is well represented by naturally occurring enzymes of potential general interest in organic synthesis. The latter reaction has precedent in the field of catalytic antibodies[11] but the former is potentially novel.

There are many current limitations in our ability to produce catalytic antibodies not least of which is the difficulty of designing and synthesising appropriate transition state analogues. Screening for catalytic activity is also a major problem. Consequently, the antibodies that are studied are usually not optimal for their purpose; the optimisation of catalytic activity is a question for current and future studies. Our experience has also shown that unintentional activities can also turn up. For example, to address aromatic substitution, we selected a hapten that would represent a positively charged intermediate on the reaction profile (Figure 3); the intention was that the positive charge would result in the formation of a complementary anion at the active site of the antibody which would stabilise a positively charged intermediate electrostatically. However, in the isolated antibodies, we did not obtain cyclisation and substitution but hydrolysis of the active ester with a spectacular rate enhancement of 3 x 10^6, one of the largest discovered. This result was rationalised on the basis of mechanistic studies with the antibody in terms of the induction of a general base in the antibody by the positively charged hapten[9] (Figure 3). It is also worth noting in the context of locating good catalytic antibodies that we also obtained an antibody that bound the hapten and analogues strongly but was non-catalytic from immunisation with the same hapten.

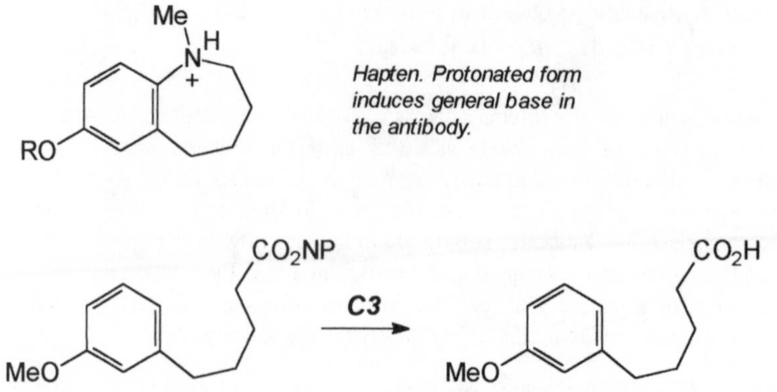

Hapten. Protonated form
induces general base in
the antibody.

Figure 3. Hapten for raising antibody C3 and reaction catalysed.

Many antibodies have also been raised in our laboratories to catalyse Diels Alder cycloadditions. One of these only has been studied in some detail. It also catalyses an unexpected reaction, the hydrolysis of an acetate ester of the product[10] (Figure 4). This antibody was raised to a hapten that is a product analogue and showed modest affinity for the substrate N-ethylmaleimide (K_m 5 mM) and a rate enhancement over the pseudo first order rate of 1,700 with 1-acetoxybutadiene as the cosubstrate. More extended kinetics have not been carried out because of the unexpected hydrolysis reaction. In order to avoid this, analogues of the original diene substrate, 1-acetoxybutadiene, have been investigated and we have been able to show that 1-aminobutadiene carbamates and 1-acetoxyhexa-2,4-diene are also substrates for the Diels Alder reaction; products were isolated and no hydrolysis was observed. The rate enhancements observed in these cases are similar to that for acetoxybutadiene as substrate. The chirality of these stable products is now under investigation. In contrast, 1-methoxybutadiene, 2-ethoxybutadiene, 1,4-diphenylbutadiene, and penta-1,3-diene were not substrates.

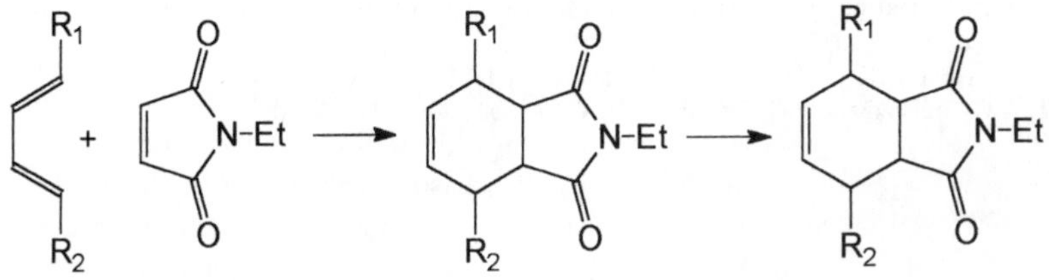

Figure 4. Reactions catalysed by antibody H11. **1, 2** R_1 = H, Me, R_2 = OAc, $NHCO_2Et$, CH_2OAc; **3** R_1 = H, R_2 = OH.

Antibodies are large proteins (150,000 D) and we have shown that a smaller fragment (Fab) of both the Diels Alder catalytic antibody and the adventitious esterase antibody are catalytically competent. In collaboration with colleagues at the University of York, we have obtained hundred milligram quantities of the native Diels Alder antibody; structural studies are now in progress. The genes for the Diels Alder antibody have also been cloned and the protein sequence is now beginning to emerge. This will allow us to plan site directed mutagenesis to improve catalytic activity and to modify substrate selectivity.

Although our studies of these antibodies are not yet complete, we can draw some conclusions with respect to clean synthesis. Antibodies with some substrate selectivity and tolerance can be obtained to catalyse reactions of potential in synthesis. Manageable fragments (Fab) are catalytically competent and products can be isolated from aqueous solution. With the experience of 25 years' work developing enzymes for organic synthesis, rapid progress in harnessing antibodies to synthesis and other tasks requiring selective catalysis can be anticipated from the academic viewpoint. As always, however, their potential use will depend upon the characteristics of the application in hand.

REFERENCES

1.   C.J. Suckling, P.J. Halling, R.C. Kirkwood, and G. Bell, *Clean Synthesis of Effects Chemicals*, AFRC/SERC Clean Technology Unit, 1992.
2.   R. Breslow, *Chem.Soc.Reviews*, 1972, **1**, 553.
3.   S.O. Onyiriuka, C.J. Suckling, and A.A. Wilson, *J.Chem.Soc.,Perkin Trans.2*, 1983, 1103; S.O. Onyiriuka and C.J. Suckling, *J.Org.Chem.*, 1986, **51**, 1900.
4.   H. Pervez, S.O. Onyiriuka, L. Rees, J.R. Rooney, and C.J. Suckling, *Tetrahedron*, 1988, **44**, 4555.
5.   G. Casy, T.V. Lee, H. Lovell, B.J. Nichols, R.B. Sessions, and J.J. Holbrook, *J.Chem.Soc.,Chem.Commun.*, 1992, 924.
6.   E.T. Kaiser, *Angew.Chem.Int.Edn.Engl.*, 1988, **27**, 913.
7.   D.J. Aitken, R. Alijah, S.O. Onyiriuka, C.J. Suckling, H.C.S. Wood, and L. Zhu, *J.Chem.Soc.Perkin Trans.1*, 1993, 597.
8.   C.J. Suckling and L. Zhu, *Bioorg.Med.Chem.Lett.*, 1993, **3**, 531.
9.   A.I. Khalaf, G.R. Proctor, C.J. Suckling, L.H. Bence, J.I. Irvine, and W.H. Stimson, *J.Chem.Soc.,Perkin Trans.1*, 1992, 1475.

10.   C.J. Suckling,   M.C.Tedford,   L.M.   Bence,   J.I.   Irvine,   and   W.H.   Stimson, *Bioorg.Med.Chem.Lett,* 1992, **2**, 49.
11.   D. Hilvert, K.W. Hill, K.D. Nared, and M.T-M. Auditor, *J.Am.Chem.Soc.,* 1989, **111**, 9261; A.C. Braisted and P.G. Schultz, *J.Am.Chem.Soc.,* 1990, **112**, 7430.

Subject Index